Cosmology in a Nutshell
(A Guide to the Nature of Reality)

ISBN 978-1-4709-4366-0

Foreword

There are statements and ideas in this book which are extreme simplifications, to say the least. Please bear in mind that this is not a book for cosmologists or scientists, it is for someone who knows little of cosmology, perhaps did science at school, may have seen media items on things like the Large Hadron Collider and wondered what it is all about. For such a person I hope this book is helpful. It will certainly help in understanding the significance of anything the LHC discovers.

Contents

List of Diagrams

Acknowledgements

The quotation at the end of Chapter 1 is from Bill Bryson's excellent book "A Short History of Nearly Everything" published by Doubleday. Some other references are from the same source.

The Periodic Table of the Elements in Chapter 5 is from "The Little Book of the Big Bang" by Craig Hogan published by Copernicus (USA). The Standard Model of Particles in Chapter 6 is from the same source.

Cosmology in a Nutshell
(A Guide to the Nature of Reality)

Introduction

Cosmology, the most fundamental of all sciences, and "the greatest adventure of the human mind", need not be difficult. All it needs is a little Lateral Thinking. Of course the biggest question in cosmology is "why does the universe exist"? Well, no one can answer that, except perhaps theologians, but we can have a look at how it came into being and how it works. The book also touches on all the various theories concerning the multi-universe, without being strongly partisan to any particular one. Time and research will undoubtedly point the way. And as far as the Nature of Reality is concerned, in this book we will ultimately arrive at partial answers, if nothing else. Even from discoveries made by the Large Hadron Collider (LHC), the most ambitious experiment of all time, it's possible that we will only discover deeper mysteries. But let's look at our attempt to understand, as far as possible, reality, the universe and everything.

Relevant diagrams can be found at the end of each chapter. The order of chapters represents a logical approach to our subject, but it may not be suitable for everyone. In the fifth and sixth chapters for instance we cover some physical concepts which, if your knowledge of physics is close to non-existent, could be difficult. It's quite possible therefore to skip on to Chapter 7. What you would then have is an understanding (hopefully) of the macro-universe i.e. the big concepts, but not of the micro-universe i.e. what goes on inside the atom. But that's alright, although I should point out that both areas of knowledge are equally mind-boggling. Perhaps just regard the whole thing as an interesting read, a bit challenging in the Lateral Thinking parts, but mainly, and more importantly, worth the time you spent on it.

2

Chapter 1
Our Universe

Our Locality

The Milky Way is a spiral galaxy, a disc of stars with, as it were, its arms bent back by the effect of the rotation of the galaxy. It's vaguely like a whirling starfish with a big body and short arms, except it takes an enormously long time to rotate even once. When we look up we see the disc from one side. It is 100,000 or 100×10^3 light years (ly) across, a light year being the distance light travels in a year, but I wouldn't bother trying to work it out. It is obviously a truly colossal distance. The Milky Way is 9×10^9 or 9 billion years old and rotates every 250×10^3 years. It has about 200 billion or more stars, bigger than most galaxies, and one of those stars is our sun. The Solar System is towards the end of one arm, but its dimensions are nothing like most of the illustrations you see.

Imagine the sun as the size of a football, then the earth is a pea 60m away and all four rocky inner planets are within 100m. An astronomical unit (AU) is the distance from the sun to the earth, which is 150×10^6 or 150 million km, or 8.3 light minutes. Next comes the asteroid belt where meteors originate. Some asteroids contain fused metals, which seems to indicate that very early in the history of the Solar System great heat was present, probably from a nearby supernova, which probably provided some raw material for the formation of the Solar System. Then at 300m comes Jupiter, with all its moons, and then the three other gas giants, Saturn, Uranus and Neptune with their moons (the acronym SUN helps remember their order). The planets are, of course, named after Roman Gods. Next is the Kuiper Belt of very large asteroids, made mainly of ice, with some traces of organic materials, including Pluto (now classed as a minor or dwarf planet) which is 40 AU away from earth. At present it would take at least ten years for man to reach Pluto. Now the sun looks very small. Much further away comes the Oort Cloud at 50×10^3 AU and 2 ly across. This is where comets exist, made mainly of ice, but also including some organic material, moving in vast elliptical orbits round the sun.

The whole lot is approximately 4.5×10^9 or 4.5 billion years old, the same sort of age as the earth, and is the result of the capture of interstellar material, including the debris from a supernova, by the gravity (G) of the sun, which is a billion years or so older. After the capture came a process of orbiting and coalescing or agglomeration into planets, moons and asteroids, via the odd collision or two. The planets rotate round the sun in the same plane, in the direction of the sun's spin and at a speed or rotational velocity dependent on distance from the sun, Mercury being the fastest and Neptune the slowest. The rocky inner planets are such because they are composed of the heavier elements which were drawn closer to the sun by the sun's G. The sun is a billion years or so older than the planets, and has a long way to go yet. About another five billion years, when, having run out of fuel, it will become a red giant. At this time mankind, if it still exists, will have to leave our earth for good and seek an alien home.

We are very lucky in that our earth has a near circular orbit, as do all the other planets, rather than an elliptical one, so our climate doesn't have the enormous extremes of hot and cold that a strongly elliptical orbit would entail. We are also in the so-called Goldilocks zone, where the temperature is just right for the water required for life to exist, but probably many other sun systems have such a planet, possibly of a similar size and gravity. The 23 degree tilt of the earth's spin axis to the plane of its orbit produces the seasons of the northern and southern hemispheres. It was the result of an ancient collision, which also created the moon.

Any mission to Mars would be quite dangerous due to solar wind radiation from the sun, which is mostly high energy, very penetrating helium nuclei (alpha particles) and which, over the 6 months or more necessary and without highly efficient shielding, could cook the astronauts nicely.

Leaving the Solar System we come to our nearest star, Proxima Centauri, in the Alpha Centauri group. On the same scale this is 1600 km or 1000 miles (in fact 4.3 ly) away, something like from here to Rome. It would take around 2500 years to reach, with present technology. After that is Sirius, the brightest star, another 4.6 ly away, which is still very close in astronomical terms. The individual stars we can see are all within our galaxy. Globular Clusters are groups of stars and there are many such groups in a galaxy. The Pole Star lies on the axis of the earth's rotation, therefore the whole panoply of stars appears to rotate around it as the earth rotates. There is a similar point in the star panoply of the southern hemisphere.

This model of our locality could be reduced by a factor of 10, in which case the sun becomes a tennis ball, the earth a grain of rice 6m away, and all distances are reduced by a factor of 10, so Proxima Centauri is about 160km or 100 miles away.

The nearest galaxy in our galaxy cluster (galaxies tend to cluster because of the influence of G) is Andromeda, which is 2.2 million ly away. It is predicted to merge with the Milky Way at some far distant time. Galaxies rotate and galaxy clusters rotate, though very slowly to our eyes. Galaxies tend to be either elliptical or spiral. They are thought to have very large black holes at their centres. The system of measurement used here is the parsec, which is 3.26 ly. There are about 10^{11} (a hundred billion) stars in an average galaxy, give or take a few hundred million, and about 10^{11} galaxies visible to us. 10^{11} is said to be about the number of peas it would take to fill the Albert Hall, or roughly the number of neurons in the human brain, and 10^{11} x 10^{11} i.e. 10^{22}, which is the number of stars in the visible universe, is said to be about the number of grains of sand on the earth, a fairly ludicrous assumption, but you can accept that it's a lot of stars.

The process leading to the black holes at the centres of galaxies was initiated by rents in the fabric of space-time after the inflationary part of the Big Bang (see Chapter 2).

Stars were initially formed from random (and we'll be coming across this word quite a lot) inconsistencies or perturbations in the amorphous clouds of H and He formed after inflation. Once an agglomeration started, G built up, pulling in more and more matter until a proto-star formed, and also proto-galaxies and proto-galaxy clusters. The high pressure caused by G raised the temperature in the proto-stars to such an extent that fusion commenced. Star formation still goes on in the clouds of interstellar gas called nebulae.

Matter ejected by supernovae formed the basis for the formation of planets round stars, again by agglomeration under the influence of G.

Because of the vast distances involved the possibility of aliens landing is unlikely unless, like Star Trek, they master Warp Drive (which is imaginary of course, although who knows?). But the likelihood of aliens existing is quite high, since there are billions of planets around billions of stars.

The Lives of Stars

Many stars are roughly the size of our sun, and about half in our galaxy come in binary pairs. The basic energy-releasing process within the plasma of hydrogen (H) nuclei in the core of a star is the fusion of several H nuclei into a helium (He) nucleus under vast pressure and temperature, releasing protons (alpha particles), and a lot of fusion energy, mainly gamma radiation, coming from actual loss of mass. The star will remain stable whilst G opposes and balances the outward explosive expansion of the fusion process, but all stars eventually run out of H and He which will have all been fused into iron (Fe), and other elements, and then most swell to become red giants like the Pole Star. Instability follows and the star may throw off matter to form what is called a planetary nebula, consisting mainly of carbon and oxygen, or if big enough it may become a huge supernova explosion, throwing off matter, gamma rays, cosmic rays (which may be anything) and neutrinos. Such a supernova if big enough may shrink back under the influence of its G to become a neutron star, which may be very small indeed, even smaller than the earth. A neutron star if big enough may shrink under its own tremendous G to become a small black hole (see Diagram no.1).

To understand this latter process you must remember that matter is almost entirely space, so when G becomes so intense that atoms break down and electrons are forced into the nucleus, to combine with protons and become neutrons (a proton plus an electron makes a neutron) and even the neutrons are forced together, the resulting mass is so dense that a spoonful of a neutron star would, on the earth, weigh roughly 60×10^9kg. This is like shrinking the earth down to the size of a pea. Lateral Thinking, as so often, is required to comprehend this. Unlike a normal star, the tremendous G pressure in a neutron star does not produce great heat or light as there are no heat- or light-producing mechanisms left, i.e. no moving electrons and no atomic fields….all that has been mostly stripped away. The possible subsequent collapse to a black hole, when absolutely no heat or light at all is present, we look at shortly.

Pulsars are basically rotating neutron stars. They emit regular pulses as they rotate and caused great excitement when they were first detected, as it was thought they were emitting intelligent signals. Quasars (quasi-stellar objects) are very bright energetic gas cloud nebulae around some galactic black holes, and possibly spiralling down into them. The brightness is due to the heat from the friction between gas molecules….the hotter, the brighter. The word nebula is the name for any amorphous object seen by telescope such as a gas cloud, which is mainly H, or the remains of a supernova, such as the Crab Nebula, which exploded in 1572, according to Chinese records, and now has a neutron star at its centre.

There is a lot of cosmic dust in nebulae within galaxies, the remnants of supernovae explosions, which occur about once in every fifty years in our galaxy. Within the nebulae can be found all the elements, molecules and even amino-acids, (semi-organic chemistry), the building blocks of proteins, which we're made of. This could have been the origin of life on earth, and certainly increases the possibility of alien life. The nebulae have become the birthplaces of new stars, and also of their planets - a continuous recycling of matter.

A white dwarf is an ordinary burnt out star. The bigger the star, the hotter it burns and the faster it dies. Stars range from cool red dwarves, which are the majority of stars since they last a very long time, to hot blue giants. Our yellow sun is mid-way between these two types of star. Giant, young, short-lived blue stars inhabit the outer reaches of galaxies, with in general more, smaller, older, more yellow or redder stars nearer the centre being drawn towards, and orbiting rapidly round, the black hole. The greater heat of the young blue stars causes electrons and thus the emitted photons to vibrate faster, which means the emitted light is higher in frequency (and energy) i.e. towards the blue end of the visible light spectrum. Blue stars may be big enough to become supernovae when they die.

The star Betelgeuse is a red giant not very far away which may well go supernova within a few generations, when it will be a spectacular sight in the night sky.

Black Holes

A galactic spinning black hole is the most awesome and sinister phenomenon that has ever existed. It is stupefying. It is G gone mad. More prosaically it's a bit like a plughole. Massive G is pulling in and crushing matter and space until it becomes pure energy, which is then "crushed" into a singularity, a thing beyond our physics and our understanding. You cannot put a substance to pure energy, so presumably it can be "crushed" down to a singularity. I write "crushed" because it's the closest word we have for a process that we cannot know. The singularity itself is similarly something that we cannot know. However there are some things we can know. There are different types of black hole, the main ones being the massive ones at the centres of galaxies, capable of pulling apart suns and planets before swallowing them, and the smaller ones being collapsed large stars. A galactic black hole probably starts as a very small quantum rent in the fabric of space-time in the very early dense universe existing after inflation, drawing in the energy in space and emerging matter, growing in size and thus developing powerful G, causing emerging stars to orbit around it and thence leading to the formation of a galaxy, which feeds the black hole. The bigger the galaxy, the bigger the black hole. Eventually G becomes infinite and spawns a singularity. All black holes have an event horizon around them from beyond which nothing, including light, can escape because of the immense G, and also because photons are crushed out of existence. The immense G causes a galaxy to orbit around its black hole, and over time parts of the galaxy are drawn down into it. When that happens it is called an active black hole. It seems that some black holes do not spin, in which case matter being drawn towards such a black hole would not spiral down into it, but would go straight down the plughole as it were, just as bathwater does on the equator, where there is no Coriolis Effect (see below).

An active galactic black hole has a torus (donut shape) of gas around it, a quasar, with jets of gas shooting out in both directions through the centre of the torus. The presence of a quasar around an active galactic black hole means that the black hole actually shines massively brightly.

At a black hole singularity T = 0, i.e. time ceases, if you can imagine that, and G is infinite (which is why T = 0, because infinite G slows T to 0, see Chapter 3 on The General Theory of Relativity). The temperature is 0^0 Kelvin (absolute zero when, incidentally, electrons, if they existed, would not vibrate and therefore would have no heat) and entropy is infinite. All these factors are opposite to those of the Big Bang. This gives us a clue to the theory that, when a galactic black hole singularity forms, a metamorphosis takes place such that on its "other side" a Big Bang occurs and keeps on being fed with energy (see Dark Energy later in this chapter), in other words a universe of some sort. Speculative but there is a mathematical proof of this, for what it's worth. Effectively T approaches 0 and G approaches infinity even at the event horizon, so that any object falling in would be stretched out and would leave a static (due T approaching 0) and fading image of itself at the event horizon.

Another way of looking at it is to say that in a singularity the overall positive energy of matter cancels with the negative energy of G in the opposite way to when they emerged and separated in the Big Bang. Again this might be why the temperature is zero. Perhaps it's better to say they combine rather than cancel, and then emerge as a Big Bang, and thereafter as Dark Energy in the subsequent universe. Well, it all must go somewhere and certainly it doesn't go into a septic tank. Note that such an emerging universe may well have different physics, even different dimensions, and may in fact not be able to survive in any significant way.

A black hole is a massive rent in the fabric of space-time. It actually pulls in space itself because space is Something not nothing, as we see later. Curious things called wormholes leading, it's supposed, to other parts of space may exist nearby, like subsidiary eddies, although wormholes are purely speculative. They could be due to the Coriolis Effect of the black hole's spin, much as the earth's spin causes the swirling of water down a plughole, or the swirl of air in a hurricane. This could lead to a catastrophic break down in the fabric of space, and a wormhole could appear. The Milky Way black hole is called Sagittarius A, it is 27 thousand ly away from us, and does not have a quasar round it.

The "Shape" of the Universe

More Lateral Thinking required I'm afraid. In fact you will need to keep thinking laterally for virtually the rest of the book. Imagine a balloon. This sort of imaginary analogy is called a thought experiment and can be quite helpful (provided of course, you can think laterally). Now take the surface of our balloon and imagine umpteen billion galaxies spaced out over it (in clusters). Now imagine the balloon expanding and the surface stretching. All the clusters are moving apart from each other. The further away a cluster is from any point, the faster it is receding from that point (in the real universe the clusters don't expand internally because of their composite G). Finally imagine the 2D surface of our balloon as actually filling the whole of the balloon. Or, more simply but less accurately, imagine a spherical cake with currants in it (clusters) being baked and expanding. Now take in the fact that in there somewhere, so small as to be invisible, utterly and totally insignificant, there is the earth. Imagine (there it is again) that from the earth you start on a journey to the edge of the universe. Bad luck, there is no edge, no "outside", nor is there a centre. You will, in fact, after untold aeons, return to your starting point, just as you would if you went round the surface of the balloon. This is because space-time (or distance-time) is a continuous curve, like the surface of our balloon, due to the influence of G (see Chapter 3 on The General Theory of Relativity). This influence is the G pull of all the matter in the universe. Nor is the universe necessarily a spherical concept. It could be saddle-shaped. How odd is that? A saddle-shape is a sort of negative curvature, whatever that is, a sort of unfolded sphere. Following the curves of a saddle, you'd still return to your starting point. Straight lines are not welcome in our universe, everything is curved.

The usual definition for this very strange thing, our universe, is that it is finite but unbounded. That naturally makes the whole thing crystal clear, doesn't it? And don't forget it's expanding at an amazing rate, and now we know that this expansion is accelerating. It's better to call it recession, rather than expansion. It's just that everything is moving away from everything else, but not from any specific centre. Our 3D space is stretching, in fact. We'll just think of normal expansion, it'll work for us just as well.

We can see as far as what is called the visual or De Sitter horizon, which is where the light we see originated some time after the Big Bang, about 1.82 billion years after in fact. (You really have to concentrate to follow this part). The age of the universe is thought to be about 13.82 billion years. This time, 1.82 billion years after the Big Bang, was when galaxies were beginning to form and the balloon was much smaller. The light from it has just had time to reach us through a space that has been expanding massively for a further 12 billion years, giving the light much further to travel, and has been red-shifted by the Doppler Effect. This is an effect which we needn't worry about in any detail that is due to the vast average speed of expansion, which stretches the wavelengths of the light radiation and thus lowers the frequencies towards the red end of the light spectrum (see Chapter 5). It's also been bent by the curvature of space. So what we see at the De Sitter horizon are proto-galaxies (galaxies and the stars in them just beginning to form). For a while, as time goes on, we will see a little more being revealed, but soon the proto-galaxies that we see will be vanishing over the De Sitter as the speed of recession accelerates, such that the light from these proto-galaxies will then never reach us.

Of course for some other being elsewhere in the universe we could be inside, on, or even beyond, their De Sitter horizon. See if this is made any clearer by Diagram no.2.

There is undoubtedly a vast region beyond the De Sitter relating to the time earlier than 1.82 billion years after the Big Bang. It's called the mega-universe.

Permeating the universe there is thought to be Dark Energy which fuels the expansion. Perhaps, I only say perhaps, this comes into our universe because we are in fact on the other side of a really huge black hole in a much bigger universe than ours.

The "Extent" of the Universe

The age of our universe is thought to be about 13.82 billion years. At present we are seeing as far as about 12 billion ly away. That's where the De Sitter is. That is, light left this region 12 billion years ago, when galaxies were forming, as we have said, and the formation of galaxies is what we are seeing there. That was 1.82 billion years after the Big Bang. And here comes another bit of Lateral Thinking. At present this region is over 40 billion ly away, the present mega-universe, way beyond the De Sitter, which is actually closer to us. This is because of the time light takes to reach us, and because of the accelerating expansion during that time. It will have developed into black holes, mature galaxies and galaxy clusters. So if you've grasped this you will realise that the present picture of the universe, that is, extending our present time to the whole universe, is not as we see it, it is smaller, because the De Sitter is closer to us. It follows that when we look up into intergalactic space we are also looking back in time.

Let's return to the balloon analogy. Imagine the tiny earth somewhere inside the expanding balloon, with a largish imaginary sphere around it, representing the De Sitter horizon, except that as the balloon expands, the sphere doesn't, it actually contracts. Someone somewhere else in the universe, if there is such a being, would have their own De Sitter sphere, which would probably intersect with ours. But the time it takes for light to travel in effect gives two pictures of the balloon, one as we see it (and also the region beyond the De Sitter), and the present one, a much larger balloon. Do these pictures help?

You no doubt realise that our universe is unimaginably (and I emphasise this) vast. Many things, from the dimensions of the minutiae of matter to the size and content of the universe are difficult to imagine. That's why sometimes ideas that begin "Imagine….." i.e. thought experiments, can be so useful. And while we're here, you realise that we and specifically our brains are products of and inside this particular universe, so it's no wonder that we find it difficult to stand apart and look on, as it were. In fact there is a body of thought which says we can never know the complete picture because we are ourselves products of what we are trying to solve. But don't let that discourage you!

I borrow the following thought experiment from the source mentioned at the beginning of the book. Imagine the Empire State Building. We are on the 100th floor. The most distant galaxies we see are on the 60th, and the proto-galaxies (the De Sitter) are on about the 20th. If you're falling head first past these floors you'll soon find out where the radiation shortly after the Big Bang (the CMB) came from.

The Cosmic Microwave Background radiation (CMB), discovered in a rather farcical way by two chaps manning a horn antenna in the USA, who thought that the hiss they heard might be due to bird droppings, is in fact the very earliest light created just after the Big Bang but stretched by the huge red shift it's undergone, so that now it is just a much lower frequency microwave "hiss". It spread throughout the universe at the time it was created and has been circulating ever since. It would have come from about half an inch off the lobby floor of our Empire State Building. If you would actually like to see the birth of the universe, switch on your TV. The white noise you see has some of the CMB in it.

Dark Matter

Because the total G of the matter within galaxies has been calculated to be insufficient to hold them together, even allowing for the massive black holes thought to be at their centres, it is thought that Dark Matter must exist within them. There are conjectured to be three main possibilities, named of course with that whimsical humour beloved of cosmologists.

Firstly there are WIMPS, Weakly Interacting Massive Particles, which should appear somewhere in the super-symmetric Standard Model of particles when it is finally completed. They are thought to be relics of the matter formation era, and now form a spherical halo around galaxies. A particle called a neutralino is the most probable possibility. Note that massive here means highly energetic, not big, since mass equates to energy at micro-dimensions.

Secondly there are MACHOS, Massive Compact Halo Objects, and here massive does mean big, which are basically dark or dead stars of some sort. More whimsical humour is evident in summing these up as DUNNOS, Dark Unknown Non-reflecting Non-detectable Objects Somewhere.

The third possibility is simply a known fundamental particle called the electron neutrino or just neutrino, which might permeate galaxies and is virtually undetectable because it has very little mass.

A fourth remote possibility is that G may vary in different parts of a galaxy or in different parts of the universe.

A more recent possibility is Dark Gas, which is simply inter-galactic H.

Whatever dark matter is, it accounts for about 27% of all the matter in the universe.

Dark Energy

Here's a Lateral Thought. Imagine you can wrap up a cubic foot of space inside brown paper – a parcel of space. Imagine you evacuate all the air. Now you give this as a gift to your nearest and dearest. He or she opens it and exclaims "But dearest, there's nothing here!" "Wrong!" you say, "I have given you a cubic foot of space, and it's not nothing, it is space and space is Something. I know you can't see it, but you can't see air either. It was made out of energy in the Big Bang and it can warp (as we will see) and therefore it must be Something. Moreover it's unique in that it can't be destroyed (except in black holes). In fact without space the universe, the earth and you couldn't exist. You owe your life to space. You should be extremely pleased with it!" (Please don't try this on your wife's wedding anniversary). We will come back to this in a moment.

Dark Energy is thought to exist because the universe's expansion is accelerating even allowing for the G of Dark Matter, which you would think might be slowing the expansion. It must account for a huge 68% of the apparent matter in the universe (matter or mass being equivalent to energy) apart from the 27% of Dark Matter, if the calculations are to balance. Therefore all the galaxies and nebulae represent only 5% of all the matter in the universe. It means that 68% of the energy that originated in the Big Bang didn't go into creating matter and space. What might fit this bill would be if we were, in fact, on the other side of a black hole in another bigger universe, and the energy being pumped in, which we call Dark Energy, is in fact the continuation of the energy that erupted in the Big Bang, which itself was initiated when the singularity in the black hole in the bigger universe was first formed. Dark energy could also be creating space out of energy, as in the Big Bang, thus fuelling the expansion of the universe.

The idea of our vast universe originating from a single black hole singularity in a much vaster universe may seem unimaginable, but it only requires the Lateral Thought that size is always relative. Being creatures of this universe we see it as vast, but if we allow that different physics and different dimensions can exist in bigger, or indeed, in smaller universes than ours, then it follows that there is in imaginative Reality no limit to what can exist "elsewhere". It then follows that our galactic black holes could also produce much smaller but sometimes viable universes "elsewhere", some possibly with the same physics and dimensions as ours. I use the term "dimensions" always to mean dimensions as in the sense of "three dimensions". If Dark Energy doesn't come from outside the universe, then it must be that a process inside the universe is turning vast amounts of matter or space back into energy.

Thus a more accepted idea than the above, although still bizarre, is that because space (or you could call it the fabric of space) is Something, then virtual potential energy particles and their opposite anti-particles can randomly appear and then combine and annihilate themselves, releasing some of the energy of the Big Bang which went into creating space in the first place. Something like this must be happening anyway in this view of things simply to fuel the normal stretching or recession or thinning out of space. As we have said, it's actually everything moving away from everything else. It's just that the process seems to be speeding up.

We have said above that space is Something, so we should also note that Nothing is what existed, or rather didn't exist, before the Big Bang, if you can imagine that, which you can't.

A third explanation is that the total G of the universe must represent a vast amount of potential energy, and that this is in fact linked to the presumably kinetic Dark Energy we are looking for, and also to Dark Matter. A better version of this is as follows. A vast amount of potential energy drains from our universe via galactic and stellar black holes, opposite to the Dark Energy enigma. If there is no other "place" or universe for this energy to feed into, where does it go? It must stay within the universe, re-emerging as kinetic Dark Energy.

A fourth explanation might be that since the photons of electro-magnetic (e-m) radiation (see Chapter 5) travel at the speed of light (C) and thus experience no time, and therefore no entropy, they last for ever. Also the e-m force within atoms gives rise to the e-m radiation, which is thus itself a force, although weak, as is evident in the fact that the solar wind, which includes e-m radiation, provides sufficient force to move a space vehicle having solar sails. Taken together, this means that all the e-m radiation that has ever been emitted, although dispersed (a sort of universal, and increasing, CMB) still exists within the universe, due to the bending of e-m radiation by G, and represents a considerable force of "outward" pressure. After all, all radiation is directed "outwards", and energy is never lost. You only have to feel the heat from the sun on your face to realise the vast amount of energy there is in stellar radiation. Stellar e-m is the release via fusion of some of the pure energy stored in matter. Is this Dark Energy? It is simply an aspect of overall entropy.

Yet another explanation, the fifth, might be simply that the universe is expanding at different rates in different regions.

Dark Energy is said to be the greatest challenge in cosmology. I should point out that the effects leading to the postulating of both Dark Energy and Dark Matter may be due to G being both weaker in some and stronger in other parts of the universe. C also might vary in this way, but that idea is a bit way out.

Diagram no.3 explains the comparative amounts of matter in the universe and may help your understanding.

A possible phenomenon called Dark Flow has now been discovered. It seems that a part of space containing many galaxies and roughly 1 billion ly across is moving in a particular direction. Whether this will prove significant time will tell. Could it be caused in some way by energy coming in through a black hole in a bigger universe? There is also a region known as the Void, where nothing much exists. Perhaps that is where such energy comes in.

The Possibility of Other Civilisations

Life anywhere probably requires carbon because it is a complex atom whose molecules can take many different forms, and it combines easily with other elements. Effectively life also requires atmospheric pressure, water, a source of energy such as the sun, and time. Life on earth started about 3.8 billion years ago. Consciousness and intelligence developed on earth after about half a billion years of higher life forms, which came out of the water. If, by analogy, you represent the earth as being 1 year old, then mankind, or Homo Sapiens, has been around for about the last 20 minutes, representing some 200,000 years, and civilisation for about 1 minute. Our civilization is very young, 10,000 years at the most, depending on how you define it. An aliens' planet will be found in the "Goldilocks Zone" around a sun i.e. where conditions are just right, mainly for water, just as they are for us. Altogether this scenario may not be too difficult to achieve in our vast galaxy, and is even more likely in our universe. It is virtually certain that primitive alien life forms like viruses and bacteria exist, but will life have evolved to civilisation?

Such civilisations, and we may presume there are some because of the vastness of the galaxy and the universe, could be far advanced, but of course out of reach unless they've conquered wormholes and are exploring. An exchange of e-m signals travelling at C might be possible, stretching over many, many years between transmitting and receiving, although extremely unlikely because of the dispersion of the energy in the signals. We have our SETI programme (the Search for Extra-Terrestrial Intelligence) which involves listening for signals, but there's little hope, I feel, which is probably a good thing. If we persist in killing each other, totally unable to achieve a global civilization, I don't think that adding yet another layer of complication would help matters.

Maths could be a common language. The aliens' size would largely depend on their G. They would have to have survived the obvious crises, such as nuclear fission and fusion, genetic manipulation, population explosion, the evolution of deadly viruses, the creation of artificial intelligence, which could go wrong (intelligence is basically complexity, different to consciousness, which is a mystery), loss of planetary resources, destruction of the environment, natural terrestrial disasters, impacts, a nearby supernova and so on. Their rate of time passing would be different to ours (see Chapter 4 on Local Time), but would gradually change to match ours as they approach, if they ever do. They could still be evolving, as I suppose we are.

It is said "we live in a universe whose age we can't quite compute, surrounded by stars whose distances we don't quite know, filled with matter and energy we can't identify, operating in conformance with laws whose properties we don't truly understand". Nevertheless the fact that we have consciousness and intelligence enough to attempt to understand it at all is possibly the most astonishing thing in the whole universe. Is it unique? Is it random?

Phases in the Death of a Large Star
(Pressure Cooker Simulation)

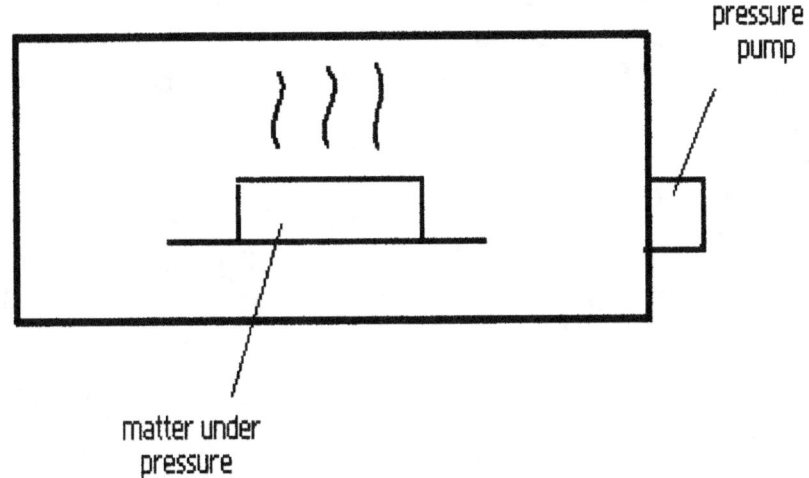

pressure
pump

matter under
pressure

Heat energy is created by the crushing of matter under strong gravity.
Each of the phase or state changes below is an example of Catastrophe Theory, the
theory of relatively sudden events.

Temperature	Phases of Matter
	Unstable period:-
$3000°$ K* (BB** 100 sec)	a plasma of nuclei and electrons. If the star is big enough a supernova may occur, which if big enough may collapse, thence
1 billion° K (BB 1m sec)	nuclei break up into separate protons and neutrons. Electrons are forced into protons to become neutrons. A neutron star is a possibility in which case the temperature collapses to close to $0°$K.
If a neutron star is big enough, pressure continues	only quarks & some leptons exist. Light does not propagate.

…… and so on

We are close to a small Black Hole when everything disappears like magic, because
positive energy is said to "cancel" negative energy (G), into a Singularity. Is this a
cosmic septic tank? No, it can be presumed it feeds another extremely small and
probably unviable universe of some sort.

* $K = Kelvin$
** $BB = Big Bang$, the opposite process to the above (see Diagram no. 4)
N.B. Leptons are a class of elementary particles including electrons and neutrinos.

The De Sitter Horizon

Earth's De Sitter is 12 billion light years away. 12 billion years is the time the light has taken to reach us, so we are seeing the galaxies and galaxy clusters being formed 1.82 billion years after the BB (when the balloon was smaller).

Anywhere 12 billion light years away from us is seeing our galaxy and galaxy clusters being formed i.e. we are on their De Sitter horizon. And of course we would be beyond the De Sitter horizons of regions which are beyond our De Sitter horizon.

EARTH'S DE SITTER

Portion of a circle, centre EARTH, radius 4.5 billion light years

SOMEWHERE

They see the Solar System being born. All stages in the history of the Solar System can be seen from different points of the universe.

4.5 billion years is the age of the Solar System

EARTH

We see SOMEWHERE as it was 4.5 billion years ago

Portion of a circle, centre SOMEWHERE, radius 4.5 billion light years

SOMEWHERE'S DE SITTER

The Cosmic Density Pyramid (to visual horizon)

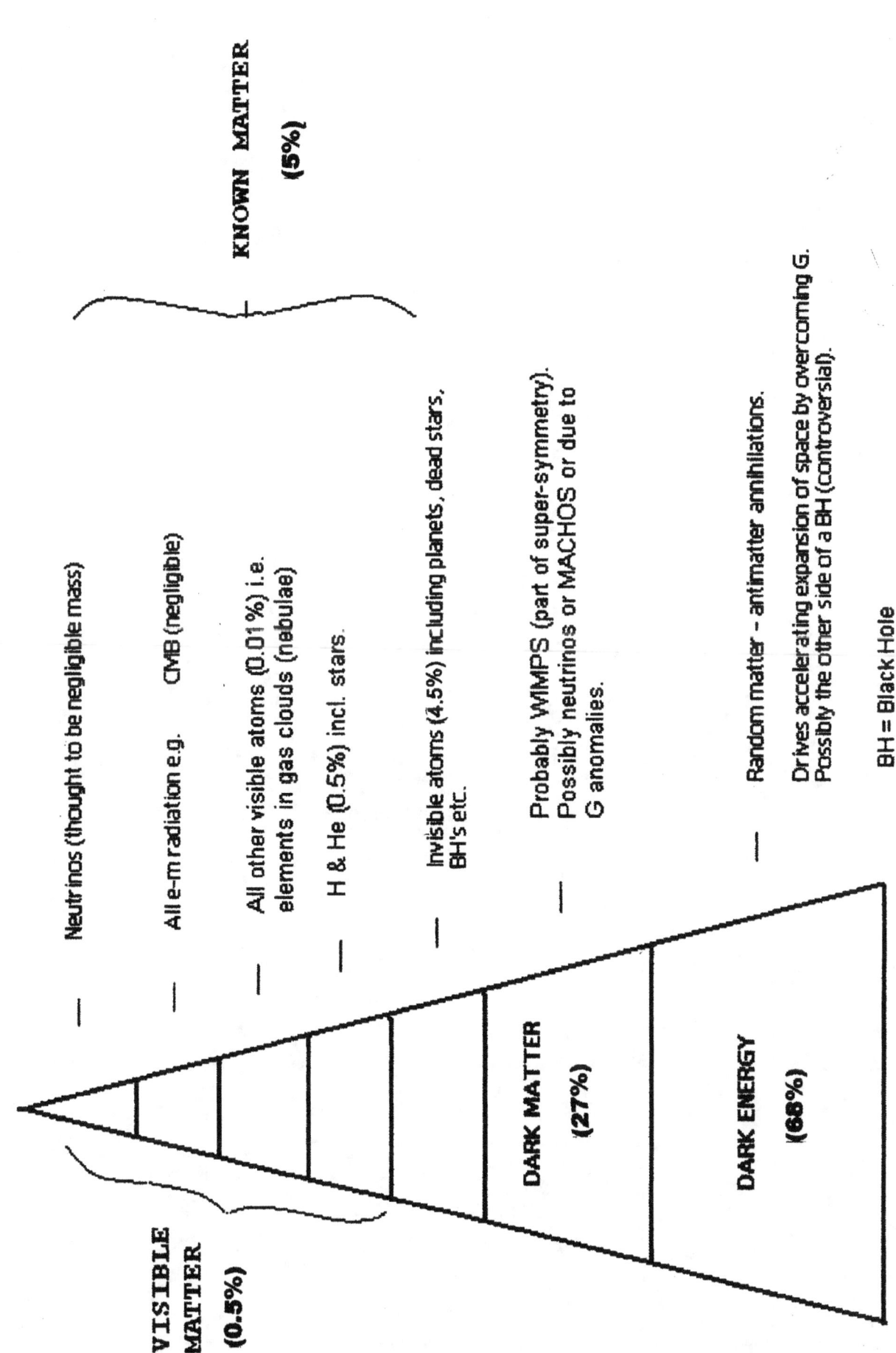

Chapter 2
Our Past & Future Universe

The Big Bang

We've already mentioned the basics. The Big Bang was a singularity….a random quantum event, possibly to do with some form of potential energy, as we have said, completely ridiculous though it might seem. We know the Big Bang was energy, and that it must have come from somewhere, so the process must have been energy changing from something unknown by us to something known by us. This is an important concept, because it begins to define what is meant by "before" the Big Bang, i.e. some unknown format of potential energy. The quantum mechanical Uncertainty Principle (see Chapter 6 on Quantum Mechanics) implies there can be an effect without an apparent cause, with a certain probability of actually happening. Our Big Bang may have been one among many, some producing viable universes, perhaps with different physics and even perhaps different dimensions and some not (or it might have been the other side of the beginning of a singularity in a black hole in a bigger universe than ours, as we have said, but let's leave that for the moment). Our minds, being products of this universe, cannot conceive of what was "before" the Big Bang. That's the only way to put it. Of course there was no "before". There was Nothing. As we have stated even space is Something. So perhaps the true definition of Nothing is simply, "before" the Big Bang. Try to imagine it.

You can't, can you? Also time doesn't exist. Try imagining that! Anyway there was a sudden random burgeoning of a colossal amount of pure energy (since it forms the whole of the universe), primordial heat, later to split into positive (kinetic) and negative (potential, mainly G)) energies, some energy going into the inevitable formation of space to give somewhere for change to occur, and therefore time to begin. It is fundamental to note that everything in the universe, or indeed in any other universe, is made of energy, including space. The burgeoning energy was vast heat without light. Creation took place in total darkness, as light did not yet exist. The heat (and motion) kinetic energy of the Big Bang was pure energy, as there were no electrons to vibrate and infra-red e-m radiation didn't exist. Entropy was zero, which means conceptually the Big Bang singularity was a point of total order. From now on, apart from intermediate formations of order, such as atoms and organisms, the overall path of entropy is towards disorder. It's why you can drop something and break it, but you can't do the opposite.

Immediately a massive inflation occurred expanding much faster than C, since light had not been created yet, and in any case it was inflation of space itself, to about the size of a football, a truly unimaginably dense ball of pure energy. Thereafter inflation slowed to less than C as heat energy was absorbed by the creation of space and matter. The G of matter also slowed it down. The temperature started dropping. Now the recession caused by inflation seems to be speeding up again due to what may be a natural process linked to the Dark Energy enigma. The inflation was inflation of the 3 dimensions making up our space, other dimensions, assuming they exist, stayed curled up. The theory of Superstrings, which we will look at later, attempts to explain this.

After inflation matter formed and the 3 forces of the atom came into being, plus G. Anti-matter was cancelled out (although might prevail in another universe, see below). The values of things like atomic charge and the values of the atomic forces, including G (see Chapter 6) probably self-adjusted to the fine balance that we have by a process of trial and error (a sort of physical evolution). If any of them were slightly different matter wouldn't form. Or you can put it down to the Anthropic Principle which says, in effect, that we're here because we're here. This provides us with little explanation and is a rather defeatist attitude, more to do with philosophers than cosmologists. There is also another Goldilocks Effect, which postulates that ours is one of a vast number of universes resulting from Big Bangs of which most fail to achieve the right physical values and never develop matter, or never develop suitable dimensions, or just collapse. Ours happens to be just right, which means that our existence is due to a lucky random occurrence, which makes us very fortunate.

After matter formed then random quantum rents in the fabric of space, with G now present, begin the formation of black holes, initiating proto-galaxies and proto-galaxy clusters. Proto-stars form within the proto-galaxies from agglomerations of matter, creating sufficient G to attract more matter and so on. Black holes mature in the centres of galaxies and develop into singularities.

Diagram no.4 shows the approximate timetable of how the Big Bang developed from the moment it started.

The End of all Things

Things move on, and now we have our present universe. The universe is still very young, but what of the far distant future? Probably recession will go on to total entropy, but there is a theory that black holes eventually disappear, and this could end Dark Energy in our universe if we are indeed the other side of a black hole. Then recession could slow down (an "open" universe), cease altogether (a "flat" universe) or reverse (a "closed" universe or the Big Crunch) if G in the universe was still sufficiently strong to cause a reverse. This could end in a singularity, which could lead on to a new emerging universe (a Big Bounce). If not, as we reach total entropy black holes may triumph and matter vanish, together then with the black holes themselves. Any of these outcomes would happen over a cosmological time so vast that even Lateral Thinking is not up to grasping it: 100,000 billion years.

From these ideas of the end of the universe, two obscure ideas of the origin of the universe emerge. The first is that our universe could itself be the result of a Big Bounce, ending in a Big Crunch, and so on, in a continuing cycle, presumably without beginning or end.

The second is more subtle. At total entropy matter vanishes and only energy in the form of photons, which have no mass, remains. Photons travel at C, at which speed time ceases, $T = 0$ (see Chapter 3 on The Special Theory of Relativity). This is also the condition at the Big Bang. The idea therefore is that all this energy has always existed and, at the end of one universe, simply exists until a random quantum perturbation occurs, drawing in energy to begin the formation of the next universe. In other words there is no Big Bang, simply a sudden random metamorphosis into inflation. It is quite opposite to the first idea, which just serves to illustrate how cosmology flounders when faced with the question of the origin of all things.

It follows from this idea that photons must be one of the main contributors to complete entropy, since they experience no time and therefore theoretically exist for ever. Thus e-m radiation including light and infra-red heat, once emitted, simply disperses throughout the universe, draining the universe of ordered energy and therefore increasing disorder, or entropy. An intriguing question here might be whether galactic black holes grow bigger over time, such that they can swallow all the photons, and once everything is gone, swallow themselves. But then where does everything go? Surely into other universes…

The laws of the Conservation of Energy and of Mass must be comparatively short term, since black holes drain everything eventually, possibly into other universes. So is there what one might call a law of the Conservation of Cosmic Energy, as universes die and are re-cycled into other universes via black holes?

This "universes via black holes" scenario possibly implies that universes would become smaller and smaller over millennia of time, until they become non-viable, when presumably everything ends.

Anyway our universe is inevitably doomed, but you needn't take out life insurance, it'll be a while yet.

Approximate Timetable of the Big Bang

Random Singularity   -   pure energy as vast heat, an explosion. Entropy 0. Space & Time start.

Inflation   -   expansion > C (which doesn't exist). G has not yet imposed itself. Inflation takes a very short time, & then slows.

Creation of Basic Particles   -   photons created then quarks & leptons, but light cannot propagate.

1 m.sec   -   protons & neutrons formed. Strong & weak nuclear forces.

100 sec   -   H nuclei form as a plasma. Universe is still dark.

370×10^3 yrs   -   H atoms form & the intense heat existing causes a quarter of the H to fuse into He, releasing vast CMB radiation. Higher elements will be created in stars as they burn up H & then He & will be distributed by supernovae. Today H & He form 98% of visible matter including a small amount of Lithium. E-m radiation & thus C come into being. Light can propagate (cf. Genesis). CMB radiation. G begins to predominate.

1 million yrs   -   proto-galaxies & stars start to form
}by G attraction

1.82 billion yrs   -   galaxies & galaxy clusters form

10 billion yrs   -   expansion begins to accelerate again due to the dark energy of space imposing itself.

13.82 billion yrs   -   present day.

Matter is merely solidified energy, the universe is energy, space is energy, energy causes change (expansion), change is time. So energy must be linked with time as well (see Cosmological Time).

The Special Theory

There's the remarkable family Stein
Gert, Ep and Ein
Gert's poems are bunk
Ep's statues are junk
And nobody understands Ein.

Einstein formulated the Special Theory using lateral thinking and thought experiments, employing very little maths. It is not as difficult to understand as it is sometimes made out to be, or as is suggested in the limerick above! It says in essence that C is a constant everywhere in space and that if you were able to travel at C (which nothing can, except photons) your time would cease. This is the key. It means that at very high speed, as you approach C, time slows significantly. $E = mC^2$ is part of the Special Theory.

Surprisingly enough, even travelling in a car your subjective time is infinitesimally slowed. But if you are travelling close to C, although your perception of time does not change, your body clock will not tick away so much. When you stop you will be younger than if you had not travelled. You would have experienced your own rate of time passing. Someone somewhere else in the cosmos travelling at a relatively faster speed would have an even slower rate of time passing, but to them it would seem normal. If you could observe them they would appear to be operating in slower motion. If they visited us they'd want to go home in double-quick time, as they'd be aging rapidly. Similarly someone somewhere else in the cosmos travelling through space at a relatively slower speed than us on our little planet, if they could see us, would see us as operating in slower motion than themselves. When referring to speed or velocity in this section, it is always relative to the cosmic frame or cosmic background.

Also as you approach C, mass approaches infinity and observed length approaches zero. The latter, you could say, is because an object, which inherently has length, cannot travel at C. The mass part is such because the energy needed to accelerate anything to C is infinite (due to inertia), and at C energy and mass would be interchangeable. It is why C is a limit. Only e-m radiation including light waves or photons can travel at C. Photons therefore, are pure energy with no effective mass.

It follows, even without any lateral thinking, that in space, where you can continuously accelerate to a vast speed, close to C, you can cover vast distances in what is, to you, a relatively short time. To someone on the earth observing, it would appear to take much longer. Let's take an extreme example. We want to visit a sun system with a likely-looking planet in our nearest galaxy Andromeda, 2.2 million ly distant. We would build the spaceship in orbit. To make the trip pleasant we could continuously accelerate for a long time at one G (G is a force of acceleration), perhaps initially utilising the sun's energy with enormous photo-electric solar sails (there is no need for streamlining in space), and then, probably using a fusion power source once beyond the Solar System, eventually reach a speed close to C. As we get nearer we would decelerate at one G, but overall that would take a very long time, many lifetimes in fact. To accelerate much more quickly (you would need to go into stasis) you would need a truly enormous power source, because the faster you go, the more difficult it would become to go any faster (inertia), the law of diminishing returns. Then because of the slowing of time (the time dilation effect) calculations suggest the trip might take just 50 years. But as perceived from earth the trip would take close to 3 million years. If you, or more likely your children, returned, it would take another 50 years, and earth would have advanced 6 million years, if it still existed. All things considered, returning would probably not be an option. It's all a bit of an exaggeration of course, however it does mean that space travel over significant distances within our galaxy (somewhat slower than this, one would think) might just be remotely possible, both for us and for aliens. Incredible but nonetheless true. This is one reason why we speak of space and time as being linked, space-time.

Your local time, or rate of time passing, as opposed to cosmological time, depends upon your cumulative rotational speed in space. If you add up all the different rotational speeds the earth is subject to, i.e. the movement of our galaxy cluster, the rotation of the Milky Way itself, the rotation of the planet round the sun (100×10^3 km/hour) and the rotation of the planet itself (negligible, 40×10^3 km in 24 hours if you're on the Equator), then you're moving relative to the cosmic frame at 370 km/sec. This is 0.0012 or 0.12% of C, and this determines the rate of time passing that we experience. The acceleration of recession might also decrease our rate of time passing. However this is not included in the calculation as recession is an inherent part of the cosmic frame or background, in other words it is actually space itself which is recessing. The effect of the earth's G on this calculation (see The General Theory of Relativity later in this chapter) is quite small as the earth's G is relatively small. Note that rotation or spinning produces a gyroscopic effect, which we might call a cosmic gyroscopic effect.

The Special Theory time dilation effect is a curious feature of that weird thing we call space-time. In this section we describe space-time from the Special Theory point of view. In the next section space-time is described from the General Theory point of view.

Before we leave the Special Theory, there is another more esoteric effect which is sometimes postulated. As we stated at the beginning of this chapter it takes infinite energy to accelerate anything to C, so anything travelling at a velocity which is a significant fraction of C has enormous energy, and thus enormous inertia. It therefore becomes very difficult to alter the velocity, and it would in fact require energy of the order of nuclear energy to slow down, just as it would to speed up – a factor to be considered in space travel. But more significantly it means velocity becomes the determining factor in what happens in space travel. Everything in space travels in curves (see The General Theory next), so it is postulated that the effect of velocity in space is to determine the steepness of the curve a spaceship will follow in getting from, say, earth to Andromeda, and thus the distance it has to cover. Light from earth to Andromeda (or vice versa) follows the flattest curve and therefore the shortest distance. The closest the spaceship can get to C, the flatter the curve it will follow and the shorter the distance it has to cover, but the greater the energy needed to speed it up and slow it down, so one encounters the law of diminishing returns. Of course the spaceship must first exceed the escape velocity of the Solar System, and then that of the Milky Way, both of which depend upon their respective G's and their relative rotational speeds through space. There is more to Relativity than at first appears. Will man ever venture beyond the Solar System and learn how to navigate in deep space? If so, what a giant leap for mankind.

The General Theory

The General Theory is extremely profound, an astonishing insight into an aspect of Reality. It is about how G affects space, and also time. G, although it can be regarded as a force or field, can also be regarded as warping space, in other words mass warps space. Thus suns and planets can be regarded as embedded in space (see Diagram no. 5). The warping of space (within galaxies) is why bodies orbit each other rather than collapsing in on each other. Also as G is a static field rather than a moving one (such as e-m radiation) then its effect is instantaneous, rather than delayed by the time a moving field would take to travel. So for instance if the sun and its G suddenly vanished, the earth would fly out of orbit instantaneously and we would all perish but, had we not, we would not have seen that the sun had vanished until 8.3 minutes later. It is worth noting, by the way, that even an object as small as a cannon ball has an extremely weak G field. We should also note that the General Theory predicts the existence of black holes.

To see why G affects, in fact slows, time, note that G is a force of acceleration, as is evident when something is dropped, and therefore G affects, in fact increases, the speed of whatever it is acting on. Therefore it will slow the time that whatever it is acting on is experiencing. Put these two together, G affecting space as well as time and noting that G pervades space, we then see why it is called space-time, the time aspect really being what we have called local time. Overall the effect of G is to slow the rate of time passing of anyone travelling through gravitational space.

Because G warps space, anyone and anything (galaxies, stars, planets) travelling through gravitational space will travel in a curve (see Diagram no.5). This is made obvious when you throw a ball into earth's G field to someone some way away. It follows a parabolic curve, which in effect is due to the warping of space by G.

Anything in orbit is negating the effect of G, a force of acceleration, so we can deduce that travelling in a curve such as an orbit is equivalent to an acceleration effectively cancelling out the acceleration of G. Anyone aboard would be weightless or in freefall as they would not feel the acceleration of G. Incidentally an aircraft travelling in a parabolic curve, just like the ball thrown in a previous paragraph, will also cancel out the acceleration of G, and will be weightless, as will be its passengers, and as indeed will be the ball.

Let's explore the idea of orbiting a little further, as it's interesting. The height of an orbit is determined by the velocity (velocity means both speed and direction) with which an object enters the earth's G field, either after having been fired upwards at an angle or coming downwards at an angle, the direction of the orbit being normally in the same direction as the earth's rotation. An orbit is normally elliptical, which at one end of the ellipse is closest to the earth and at the other end is furthest away from the earth. A rocket's engine can be fired to change an elliptical orbit to a circular one. Velocity varies around an elliptical orbit, as with a ball thrown upwards in a steep parabola (which is like a part elliptical curve), which at the top of the curve almost stops altogether for a moment. Weightlessness applies all round an orbit, even an elliptical one.

The simplest way to consider the phenomenon of orbiting is as follows. The centrifugal outward force of an orbiting object, travelling in a curve and therefore undergoing an effective acceleration, is in effect cancelling out the centripetal inward force of acceleration of G. The centrifugal force makes the object weightless, and the centripetal force of G holds it in orbit. This is like a boy whirling a conker on the end of a string, where the string represents G.

Note that the cancelling out of the centripetal force of the earth's G by the centrifugal force of the rotational or angular velocity of an orbiting object is a concept independent of the direction of the earth's rotation, and this implies that an object can orbit in any direction round the earth, even in opposition to the earth's rotation.

Objects in lower orbits must orbit faster than objects in higher orbits because G is stronger close to the earth. An object fired upwards at a velocity greater than what is called the escape velocity will leave the earth's G field behind and escape into space. A meteor is an asteroid travelling with the wrong velocity to achieve an orbit. It will therefore impact the earth or burn up in the atmosphere.

At a height of about 300km lies the vague outer edge of the main part of the ionosphere, which is virtually the edge of the earth's atmosphere. The orbit of an object below this height will decay rapidly, meaning it gets lower and lower, due to the friction of air molecules, finally falling to earth or burning up in the atmosphere. At a height of roughly 40×10^3 km (about the same as the circumference of the earth) above the equator the orbital rotational velocity required for a circular orbit at that height has slowed to where it matches the earth's rotational velocity. Thus an object, such as a communications satellite, given the right initial velocity to achieve that orbit, will always remain above the same place on the earth's surface. It is called a geostationary orbit, and the height introduces a significant time delay into communications using the system, such as live TV coverage from abroad.

The deduction we can make from all this is that drifting or moving in a curve due to G anywhere in space near a mass, including an orbit, negates the effect of that G. Moving in a curve due to G is much the same thing as imagining a spaceship entering and undergoing the orbital rotation of a galaxy or solar system as it passes through. Anyone aboard a spaceship that has ceased propulsion and is therefore moving in a curve in gravitational space would be weightless and would also have no sensation of speed, mainly because there is no air in space, which also means the spaceship would not slow down, as there would be no friction with air molecules. Also of course the rate of time passing is affected, both by the fact that you're in a G field, even if you're not experiencing G, and by speed.

There exists a region somewhere between the sun and the earth, but closer to the earth, where their respective G's cancel. If you stop your spaceship there you can hang there, still weightless. It's one of the few places in the Solar System you can do this.

In a similar fashion to all this, the vast G of the sun (and the gyroscopic effect of rotation or spinning) holds the planets in their orbits, and since there is no air in space, they remain there. They are stable, or in equilibrium. If the earth is hit by a giant asteroid or a comet, it is not so much the devastation that would be the worry, but rather that the earth's orbit might be affected, possibly leading to catastrophic climate change. Mercury orbits the fastest because it is closest to the sun. Any object in space which is not under propulsion, within or near the solar system, will travel in some sort of orbit, usually elliptical, around the sun. Applying this general principle to the vastness of the galaxy explains why the galaxy rotates under the influence of the truly enormous G of the spinning black hole at its centre, as do all galaxies. Another way of looking at all this is as the warping of space by G. Galaxy clusters rotate, and probably the whole universe rotates.

Getting back to how G slows time, consider this. G on a mountain top is weaker than at the base. The base is nearer the hot magnetic iron core of the earth from which most of the earth's G derives because it is denser. Both parts of the mountain are moving relative to the cosmic frame at virtually the same speed. But a clock on the mountain top will tick a little faster because it has a weaker G acting on it. Odd but true. So G is warping time as well as space. The space-time warp at the base of the mountain due to G is slightly different to that at the top. The value of G decreases quite sharply with height above the earth.

This being so, the amount by which time speeds up due to G (relative to time on the surface of the earth) for an object in a circular orbit at a certain height above the earth, such as a space station, is effectively determined by the lower value of G at that height. In practise the effect is extremely small. The total effect depends on how long an astronaut stays up there, but it is still extremely small. However it has been verified by experiment and the Geostationary Positioning System (GPS) allows for it, because time at the height of the GPS satellites runs a fraction faster than on earth. The higher the orbit, the less the G and thus the faster a clock will run compared to one on the earth's surface. However the orbiting satellite is travelling faster against the cosmic background than the earth's surface, and this slows time and thus reduces the effect, and is in accordance with the Special Theory. It is still earth's speed relative to the cosmic background that determines the rate of time passing on the earth's surface, the effect of G on that being very small.

For an object in an elliptical orbit, where G and speed vary with the varying height of the orbit, the rate of time passing will similarly vary with the varying height, but you would not notice the changes, even if they were significant.

If you achieve escape velocity and leave the earth behind altogether, carrying an impetus coming from your own orbital velocity and from the earth's orbital velocity, your rate of time passing will depend much more on speed. If you stop propulsion you will be travelling in an elliptical orbit around the sun, your rate of time passing depending mainly on your speed, but to some extent on the sun's G.

If your velocity is such that you manage to escape from the sun's G and thus from the solar system, but then cease propulsion, you would be drifting at speed (momentum) along a vast curve set by the faint G fields of the nearest stars. If you have propulsion you would be continually accelerating, possibly to a truly vast speed, thousands of kilometres a second (note that whilst accelerating you would be creating G within the ship) continually slowing your rate of time passing by a significantly measurable amount. You could live a very long time…earth time, of course. The misfortune is that you would not notice it in your time frame. However, for the sort of space travel and speed so far available to us, for instance to the moon and back, the slowing of time is very, very small.

Overall, when we say that your rate of time passing depends on your speed relative to the cosmic background, we assume that is incorporating the effect of G, which is usually by far the lesser of the two, unless you are subject to a very large G, in which case it is G that determines your rate of time passing, such as near a black hole, rather than speed.

Light is bent by G because space is bent or warped by G. So when we look into space, the perceived position of an object may be vastly offset, compared to where it actually is. This, of course, further complicates our picture of the universe, but let's not worry about it…there is enough to grasp as it is!

For completeness I should mention that if a G field is catastrophically disturbed, as in a supernova, this would cause a gravity wave, which would be a ripple in the fabric of space. There would also presumably be gravitons

If, as Einstein tells us, mass distorts the fabric of space, then G as a force may be an illusion, and although a supernova would cause ripples or waves in the fabric of space, gravitons would not exist.

Incidentally weight is mass with G acting on it. In the particular case of the Earth's G we regard them as the same thing. On the moon weight is less than mass because G is less.

Summing up, the General Theory tells us that G rules in space. Space is effectively warped, or curved, by the local G fields. Every object moving in space, and even e-m radiation, moves according to these warps or curves.

With regard to time, the Special Theory tells us that the rate of time passing for an object depends on its speed, and the General Theory tells us that G can also affect or warp the rate of time passing. The closer you are to a large mass, especially a black hole, the more your rate of time passing will depend on G rather than on speed. Hopefully you will achieve orbit, but if not, as you pass the event horizon and enter the black hole, your rate of time passing will slow to 0, although that will be the least of your worries. The reason G slows time is because anything moving due to G is moving with the equivalent of an acceleration, which has nominally to do with speed, and therefore can slow the rate of time passing. What links space and time together so it becomes space-time is G.

Special Theory (Simplified Illustration)

Path of a space ship travelling at high speed, experiencing slower time than on earth i.e. effectively slow motion as seen from A

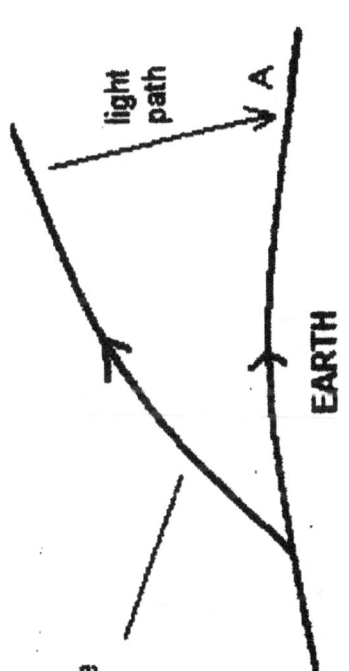

light path

A

EARTH

The curve simply indicates the space warp due to G in General Theory. Time slows for this reason also.

General Theory

the "fabric" of space

Warping of Space in 3D

Warping of Space in 2D

Anything, including light, travelling into the warped region, first begins to drop down towards the body and is then bent around it and sent off at an angle. If slow enough an object will remain in orbit, the height of the orbit depending on its velocity. If too slow it will fall into the body. In all these cases travelling in a curve is equivalent to acceleration, or rotational velocity, and thus slows time.

Chapter 4
Time

The long unmeasured pulse of time moves everything. There is nothing hidden that it cannot bring to life. Nothing once known, that may not become unknown.

Sophocles, Ajax.

Local Time

By local time we mean the rate of time passing in your cosmic locality, which is dependent on the cumulative rotational speed against the cosmic background of the planet or spacecraft you're on, and also on the strength of the G you are subject to. Someone somewhere else would have their own cumulative rotational speed and G, and would therefore have their own rate of time passing.

Were someone somewhere else to be viewing us, they would be viewing us sometime in our past, because light takes time to travel. It follows that they would have their own view of the universe, different to ours, such as their own De Sitter horizon, as well as their own time.

Moving between us and somewhere else involves the twin theories of Relativity and how they create the concept of space-time, and thus the odd things about time that would happen on such a journey through space.

Thus if someone from somewhere else were to travel to us, their rate of time passing from our point of view would speed up to match ours as their speed would be slowing and adjusting itself to ours. They would notice no change.

Cosmological Time (or Absolute Time)

Yet another lateral thought. Theoretically all stages of our past are viewable from different "elsewheres", and this applies to any place or planet in the universe. So in this sense all of past time is present in the universe. Think about that! But this lateral thought does not finish there. Whichever place we are viewing, we are viewing its past, but it also has its own present, which we will view in our future. Again this applies to everywhere in the universe, so in that sense much of the future is present as well! What I think we can say more simply is that cosmological or absolute time encompasses the whole universe.

Cosmological time is said to have an arrow, meaning it only ticks away in one "direction", that of entropy. In fact time only really exists because the Big Bang introduced change. You could say time is nature's way of stopping everything happening at once.

Putting it another way, change involves motion, and motion is linked with time, as in the Special Theory of Relativity. So there is a more fundamental aspect of Reality to be discovered here. Motion is kinetic energy, so since motion is linked with time, it follows that energy must be linked with time, as in entropy. Truly energy is everything, including space and time. It's a shame that we don't know what energy really is! Or space, for that matter.

35

Time as a dimension is just a concept. You could, if it makes it any easier, consider it as follows: there are the three dimensions, up and down, side to side, back and forth, but a fourth can be added, past and future. The flow of time is this fourth dimension.

As we have seen, cosmological time starts at a Big Bang singularity and ceases at a black hole singularity.

Time Travel

There was a young lady from Wight
Who could travel much faster than light
She travelled one day
In a relative way
And arrived on the previous night

This, along with the previous limerick, seems to be a by-product of many years of expensive research. So let us consider the possibility of time travel. Travelling to your own past is obviously not possible. However, if you could go through a wormhole, for instance to another part of the universe, you would be in the future as far as where you came from is concerned, because light takes time to travel. A more feasible way you could travel to the future would be by travelling away from your planet at vast speed and then returning. Your time clock would not have ticked away as much as a time clock on your planet. Relativity tells us this is so.

Because of something called the Uncertainty Principle (which we will explore in Chapter 6) every event at quantum level has two possible outcomes, similar to a particle being either a wave or a particle. In normal physics only one or the other is true at any one time, once seen or detected by an "observer". There is a bizarre theory that says both will be true. In other words Reality splits at this point, branching off into two slightly different parallel universes or realities. And this would occur with every "observed" quantum event, leading more or less to an infinity of parallel universes branching off from each other all the time. So, for instance, there would be umpteen of you, and umpteen more of you created virtually every instant. One would need to ask one's friends if this would be a good thing…!

If this theory were true, then theoretically a person might be able to travel back in time, although not to a time before the time machine needed was invented. If this time traveller then altered circumstances, for instance by eradicating his or her parents before his or her birth, branching universes would be created in which the traveller had never been born, avoiding the obvious paradox. Of course in the traveller's own timeline he or she would have travelled back in time, eradicated those unlucky parents, and then vanish, and his or her timeline would cease. And presumably this parallel universes idea, which is called the "Many Worlds" theory, extends throughout our Reality. It is a highly outlandish theory but it has credence with some cosmologists. They cannot see a logical alternative, not even the idea of statistical probability affecting which of two possible quantum events actually occurs, which would seem more sensible. This is discussed in Chapter 6. One would think the time machine itself could only be invented if the "Many Worlds" theory were true. Note that cosmologists tend to use the word "world" as being interchangeable with universe, for some strange reason.

Energy & Matter

All usable energy in our universe ultimately comes from the atom. The beginning, the Big Bang, was a sudden random explosion of pure energy, and was when time and space came into being. Given enough time and space, energy was converted into matter by a series of apparently random but probably inevitable small steps. Thus an atom of matter is ultimately all energy, but if we exclude the energy contained within the composite particles of an atom, then some of the natural energy easily available once stimulated is electro-magnetic (e-m) radiation, some of which is light, derived from the e-m force or field which binds the electron to the nucleus. The other way that natural energy is easily available once stimulated is the vibration of the electron, which is manifested as heat, and causes the motion of atoms. (Electrons vibrate to some extent in atoms without any outside stimulus). A fire is a fairly catastrophic example of the release of these energies.

The movement of electrons between atoms is a separate largely "man-arranged" (if you'll excuse the phrase) form of energy, and is manifested as the energy in electrical current and as the energy in chemical reactions, both of which are dealt with below. They provide the stimulus to release the natural energies outlined above.

So light and heat are the main forms of natural energy – you know that anyway from your domestic energy requirements, just as you know that electrical energy brings you your natural energy requirements, light and heat, as does chemical energy in the form of the flame.

Without the sun the earth wouldn't exist, but if it did, it would be a dead planet. So it is the sun's atomic energy that is the ultimate source of almost all of our energy here on earth, such as carbon-based fossil fuels (via photo-synthesis), and reaches us as e-m radiation, mostly light and infra-red radiated heat, generated by a process called fusion within the sun.

Energy which arises from the heat of the earth's core, such as is manifested in volcanoes and earthquakes, is also present in the earth's environment. Exceptionally the heat from volcanic fissures is the energy source for strange forms of underwater life, rather than the sun, and may have been the original source of life, which is, ultimately, concentrated energy.

You cannot hold energy in your hand, as you can with matter, so what exactly is natural energy? Nobody really knows but, as we have said, in simple terms it is e-m radiation, which includes light and radiated heat, and also the vibration of electrons, which when passed from one electron to another and so on constitutes conducted heat, as opposed to radiated heat. The vibration of electrons is why hot things expand.

Radiation is motion and vibration is motion, so any form of motion is another manifestation of energy, and motion of anything when brought to a stop creates heat vibration. No need to get confused by all this. It's simply that all these forms of energy i.e. e-m radiation, heat and motion, are interchangeable. It's what you'd expect. Energy is energy. They are collectively known as kinetic energy, which can easily be passed from one form to another.

All natural external forms of energy, like sound and wind, are simply manifestations of kinetic energy, which in these two particular cases is the energy of motion of air molecules, a process which is enabled ultimately by the sun's energy, as in thunder and the heat energy in the world's weather systems.

Man-arranged forms of energy such as electrical, mechanical (man-arranged motion) and chemical energy are also simply manifestations of kinetic energy.

To clarify electrical energy a little, when the atom was created, shortly after the Big Bang, potential energy was stored within it in the form of coupled charge (see Chapter 6), mainly electrical charge, positive on protons, negative on electrons, creating the e-m force or field which binds the electrons to the nucleus of protons (and neutrons), and to some extent controls the motion of electrons round the nucleus. So charge or the e-m force or field is potential energy. Electrical current is the flow or movement of electrons carrying charge between atoms, and is one of the most easily available man-arranged sources of energy. It can produce the natural kinetic energies of heat and light and also man-arranged motion. Of course it does exist naturally, for instance in the form of lightning.

As far as chemical energy is concerned, man-arranged chemical reactions between atoms, such as fire (which also of course occurs naturally) are another source of much of our energy, and are simply other ways of turning the potential energy in atoms into usable kinetic energy (usually heat) via the movement of electrons between atoms, and also of course to create useful chemical compounds. This also occurs naturally, as in the formation of the chemical compound water from hydrogen and oxygen. A battery is an interesting example of man-arranged chemical energy in that it leads directly to electrical energy. It's also an example of potential energy leading to kinetic energy.

It's getting a bit complicated, I'm afraid, but at least the information is here, and you can easily refer back to it whenever you want. But to sum up, the Big Bang was the sudden appearance of pure energy, creating space out of energy, and inevitably time, and soon matter itself in the particular form of the atom. The particles within the atom including electrons were created out of kinetic energy, an aspect of pure energy, and the forces or fields within the atom, holding it together, out of potential energy, similarly an aspect of pure energy. Electrons can be made to move between atoms in the form of electrical and chemical energy, especially by an outside agency like man, as well as naturally. In doing so these energies can disturb the potential energy forces or fields in the atom, producing chemical compounds or releasing the main forms of natural energy, i.e. e-m radiation and the heat vibration of electrons, and also man-arranged motion (via both electrical energy and chemical energy in the form of fire). We'll concentrate mainly on the natural forms of energy, since they are significant components of Reality.

Inside the atom, in simple terms of course, there is the nucleus with electrons going round it at different energy levels (or shells, or orbits). Now, as far as e-m energy is concerned, energy is absorbed when an electron jumps from a low energy level to a higher one i.e. further away from the nucleus, and is emitted as e-m radiation, often light, when the electron falls back from the high energy level to the lower one. It's where the light in the universe comes from. Each time an electron falls back it emits a discrete package of e-m energy, a quantum. If it is in the visible light part of the overall e-m spectrum (see Diagram no.6) it is called a photon, but for convenience we shall use the word photon for e-m radiation above light frequencies as well. If such radiation is in the infra-red part of the overall e-m spectrum, it is radiated heat, as opposed to electron vibration heat. E-m radiation below light frequencies becomes less and less photonic in nature and more obviously wavelike, although in general e-m radiation can be regarded as either a string of particles (photons) or as a wave (energy). To understand this better it helps to regard the photons as bursts of many cycles of very high frequency, very short wavelength (or cycle length), sine wave energy, which together form a continuous wave (see Diagram no.6 for a depiction of one cycle of sine wave energy).

A simple example of e-m energy production is the light bulb. Electrical current is a stream of free electrons passing from atom to atom through a material, usually a metal, whose atoms have in their outer energy level only a few of the number of electrons required to fill that shell, making it easy for them to move. However if the current is forced by voltage pressure through a material which doesn't have only a few electrons in its outer shell, in other words a resistive material, then heat is generated by the forced vibration of electrons in the material. This vibration can be passed to the electrons in the atoms of the gases which constitute air, and heat is thus conducted away, although in a light bulb there is a vacuum to prevent this. However more importantly, if the vibration due to heat is sufficiently intense, electrons can be forced up to a higher energy level (excited atoms). They will then fall back to the lower one, each releasing a photon of e-m energy radiation, having a frequency usually within the visible white light range, and often with a particular colour frequency distinct to the material or element involved (white light is made up of all the colours of the rainbow). If the element in the bulb is chosen correctly, e.g. tungsten, then the frequencies within this radiation will be in fact most of the white light part of the e-m spectrum, and possibly also the infra-red part (heat). Other types of lamp use the same general principle. Above light frequencies, by bombarding atoms with electrons, high frequency high energy e-m radiation, i.e. x-rays and gamma rays, can be produced.

When e-m energy is received it changes its state. Some frequencies are absorbed, usually as heat, or sometimes as electrical current, and some are reflected, for instance the light frequencies, with some frequencies or colours reflected more than others, depending on the reflecting surface. This is how we see things and their colours. Our eyes and brains respond only to those frequencies, and see them as colours. White light is made up of all the colours. A white surface reflects most light and heat, black absorbs most light and heat. Other eyes (and brains) in the animal kingdom can see frequencies slightly above or below the range that our eyes can see, for instance some animals, like snakes, can see infra-red.

Below light frequencies, in a transmitting aerial, electrons oscillating back and forth along the aerial can produce an e-m radiation. This is because the oscillating electrons are produced by a resonating alternating current (AC), a sine wave of current (and voltage), and the electric and magnetic fields thus produced (see below) combine to produce e-m radiation. This radiation can be made to carry information, as with radio and television. Such relatively low frequency radiation is absorbed as current in a receiving aerial.

Photo-electricity, the production of moving electrons from light energy, is an example of a particular type of e-m absorption.

Going back to electricity for a moment, electrons when moving create a magnetic field around them. Note that the electrons moving around inside atoms can be lined up by applying magnetism and this enhances the magnetic effect. This is how a magnet is made. Also electrons changing orbits are moving, and this is how the magnetic part of e-m radiation is produced. On the other hand an electric field is produced where there exists a potential (or voltage) to move electrons from one place to another. So, electric and magnetic fields are intrinsically linked together. Moving a wire through a magnetic field will cause electrons to move in the wire, otherwise known as a current. This is the generator principle, whereas moving a magnetic field across a wire carrying a current will cause the wire to move. This is the principle of the electric motor. Big coils of wire are used in practice in these machines in order to intensify the fields.

Getting back to energy, of course there is $E = mC^2$, Einstein's great equation that came out of his Special Theory of Relativity, where E = energy, m = mass and C = the speed of e-m radiation, including light. This tells us definitively that matter or mass is made of energy. It is related to the equation for kinetic energy, the energy of a moving object, which is $e = \frac{1}{2} mv^2$, where v is the speed or velocity of an object whose mass (or weight, in kilograms) is m. Because C is such a relatively large number it tells us that a little bit of matter is made of a very large amount of energy.

Fission or fusion of the nucleus are the two processes which can liberate the energy stored in the actual particles of matter, fusion being the more efficient since it involves a greater loss of mass, as in the hydrogen bomb, although this is a simplified view. The energy of the sun and stars derives from the same process, the fusion of hydrogen atoms into helium.

Then we come to C itself, the speed of e-m radiation in space. It is always constant, and nothing can go faster (it is thought). E-m energy radiates as sine waves having a frequency of F cycles per second (or Hertz) and a cycle wavelength (λ, the Greek letter lambda). Thus $F\lambda = C$…the energy and penetration of the radiation increases with the frequency. Low frequencies travel long distances, and are produced by long aerials. Higher frequencies are line of sight and are produced by a variety of devices. The value of C is almost 300 million or 300×10^6 metres per second in a vacuum, which is 300 thousand or 300×10^3 kilometres per second, about 7.5 times round the equator in a second. E-m radiation can appear as a sine wave or alternatively as apparent moving particles, photons, as we have mentioned, which can be imagined as very short bursts of sine wave energy, which together form a continuous wave. The cyclic nature of a sine wave is due to photon vibration. At higher energy levels photons vibrate faster, giving rise to higher frequencies of radiation often referred to as being "hotter" (so natural energy is largely just the vibration of the photon and the vibration of the electron, the faster and bigger the vibration the higher the energy). Normal e-m radiation is many sine waves at a particular frequency or range of frequencies. Laser light energy is crammed into one sine wave, usually in the red end of the visible light spectrum.

Potential energy is kinetic energy waiting to happen, as it were, such as can be acquired by creating a battery or by lifting an object in a gravitational field (G) then leaving it on a shelf, say. It has then acquired potential energy, which is kinetic energy of motion waiting to happen when it eventually falls. That is then converted to heat and sound when it hits the ground, though the heat part is usually not obvious. Potential energy can be regarded as a negative version of kinetic energy, in that it is stored kinetic energy i.e. it is taking some kinetic energy out of active participation in the world. Thus we can say that negative or potential energy can both absorb and produce positive or kinetic energy. Gravity or a gravitational field can be regarded as having negative or potential energy. The composite particles of an atom can be regarded as solidified positive or kinetic energy.

Consider the following simple thought experiment. It would take positive energy to pull the earth apart against the force of G. G must therefore have negative energy.

Let's consider potential energy a little more deeply and a little more laterally. Kinetic energy can be transformed into potential energy by the action of working against G, and potential energy can then be released and transformed back into kinetic energy by working with G. We don't know exactly what kinetic energy is, but at least we can see its effects. Potential energy is even more obscure – it's nothing really, just a potential. Even so, this is an important concept, as it must relate to the Nothing that "existed" before the Big Bang of kinetic (better called pure) energy. It must have been a Nothing with potential – a strange concept, but physically possible, it seems. To achieve it, must kinetic energy first have been expended in some way, as with lifting an object in the illustration above? These ideas lead to the cyclic universe theory (see Chapter 2 on The End of all Things). But as always with these theories you come up against the question of the original origin of the energy, of whatever form. The potential energy idea is all very well as potential energy is basically nothing, or Nothing, and it gets us a step closer, but we're still faced with the question of where does the potential energy come from? Theology…? A Higher Intelligence…? Or does it simply exist, and that's it? Take your pick, but to help you consider this we will explore these ideas a little more later on.

If you use a device to turn one form of energy into a more useful form, such as an engine turning heat into motion, it will never be 100% efficient - there will always be some wastage of energy into other non-useful forms, like wasted heat and sound. It's why a perpetual motion machine is impossible.

Since energy cannot be created (although we will examine that concept later) then what we were given in the Big Bang is all we have (perhaps), although by the law of the Conservation of Energy, when it changes its state, as in a fire, we don't lose any, it just passes on in a different state. Also it is always changing towards ordered states in the short term, like atoms and planets and humans, but towards a completely disordered state in the long term. This latter is the law of Entropy. So the energy in the universe, which started out in a highly ordered state of low entropy (the Big Bang, vast heat from a point source), is inevitably headed towards an ultimate completely disordered state, very spread out and very cold, a state of high entropy. In about 100,000 billion years from now, so not to worry. Entropy is why you die.

The Atom

The atom is a very curious affair. It isn't really like the sun with the planets going around it at all. Let's look at the nucleus. It's made up of protons with a positive charge and a roughly equal number of neutrons with no (i.e. neutral) charge. The number of protons determines the chemical properties of and identifies an element (the atomic number) and is equal to the number of electrons going round the nucleus. The number of protons plus neutrons is the atomic weight. Charge is just a word identifying a particular characteristic of some atomic particles, sometimes called coupled charge. Electrons have negative charge and it is the force of e-m attraction between the orbiting negative charges of the electrons and the equal numbers of positive charges of the protons which keep the atoms tightly bound and in equilibrium. It is very vaguely like the equilibrium obtained by satellites orbiting the earth, whose outward centrifugal force equals the inward centripetal force of gravity. Spin is another sort of charge. But let's get back to the nucleus.

Each proton and neutron is made of three quarks, so-called from a line in "Finnegan's Wake". Originally thought to be only two in number, it was suggested they be called "Partons" after Dolly. Physicists humour. All these nuclear particles are held together by the Strong force, which also overcomes the mutual repulsion between the positive charges of the protons (the neutrons help). It is this force that supplies most of the energy released when nuclear fission occurs. Fission is initiated by firing neutrons, having no charge, at the nuclei of an unstable element like uranium, releasing even more neutrons, and so on, a chain reaction. When the first fission experiment was conducted (in a squash court as it happens, since that is, in effect, a containing box) it was a carefully controlled critical density chain reaction, such as now happens in nuclear power stations.

When the first experimental bomb was exploded, meaning the critical density fission process was not controlled, but happened all at once, some scientists feared an uncontrolled chain reaction would occur, and the world would explode. So they did it in a desert, hoping that would reduce the risk as well as the amount of devastation, which was considerate of them. The world didn't explode, but a fearsome thing came into being.

The first atomic fission bomb dropped on Japan was based on radio-active uranium. The second, bigger and more powerful, and called Fat Boy, was based on unstable radio-active man-made plutonium.

Now let's look at dimensions and comparative sizes. An atom is unimaginably small, much, much smaller than the point of a pin, but amazingly can be seen by the most powerful electron microscope. Even so our knowledge of it is mainly by deduction from experiments and mathematical analysis. Inside the atom, although it's so small, it is nevertheless almost entirely space. Let's look inside it, and now you have to do your first bit of Lateral Thinking. First let's look into the nucleus. If you imagine the three quarks inside a proton or neutron as being about the size of tennis balls, then it is thought the proton or neutron itself is about the size of a room, filled with the strong force. Now shrink the protons and neutrons to tennis balls and the nucleus is then about the size of a room, filled with the strong force. Now shrink the nucleus to about the size of a pea, and the atom is a cathedral, filled with the e-m force and electrons the size of pinheads.

Yet the nucleus has 99.9% of the mass of an atom. The electrons are orbiting and their orbits, at different energy levels, look like wobbling smoke rings. This is because the electron like the photon can appear either as a wave or as a particle, depending on how you measure its effects, and the only way to depict it is as a probability function (see Chapter 6 on Quantum Mechanics) which can be visualised as a wobbling smoke ring, or more properly, ellipse. These orbits exist at the different energy levels mentioned above, with a set number of electrons in each level if it is complete. When you touch something, you are touching electrons.

Atoms can combine to form molecules…there are something like 45×10^{18} air molecules in a cubic centimetre of air. 10^{18} is 10 followed by 17 noughts, which is a million million million, or a billion billion. Imagine how many more there are in a solid, and yet it's virtually all space. Things appear solid mainly because the e-m force binds an atom tightly and also binds atom to atom, and this is what stops you falling through the floor. It is said that, if you could remove the binding forces, then all the actual mass particles in the bodies of the whole of the world's population would condense into the size of a sugar cube (which would be extremely heavy).

There are in fact 3 forces or fields that hold the atom together. E-m, binding the electrons to the nucleus and involving electrical charge, the Strong Force, binding all the quarks in the nucleus together, despite the fact that the positive charges on the protons repel each other (the neutrons help) and the Weak Force, which is to do with radioactivity in unstable elements like uranium. In such elements a very small amount of nuclear matter continually converts to energy and radiates away, sometimes destructively. All the man-made elements that follow uranium in the Periodic Table (see Diagram no.7a) exhibit the same phenomenon.

The three atomic fields, when static and contained within the atom, are holders of potential energy, which can be regarded as a stored or negative form of energy. When a static potential energy field is disturbed, it produces radiated kinetic energy, and kinetic energy is regarded as a positive form of energy. E-m is the main radiated field, i.e. radiated outside the atom, and e-m in particular can also be static outside the atom, usually as separate electric and magnetic fields, as in motors and generators. All static fields carry potential energy.

Just as e-m when it's disturbed and radiated can be regarded as a wave or as force-carrying particles (photons), so can the other two forces when disturbed and radiated, the particles being gluons for the Strong and particular types of bosons for the Weak, although together with photons they are all collectively known as bosons. Only e-m radiates over any distance.

There's also G, a fourth atomic field, but it's different to the others in that it's so weak inside the atom. Mainly its influence is outside the atom as a static potential energy field. And because any object, from a cannon ball to a planet, has a vast number of atoms, its G becomes quite strong. If radiated, possibly from a supernova, its particle equivalent would be the graviton, yet to be found. Of course gravitons could also be interpreted as gravitational waves.

Some of E in $E = mC^2$ comes from the enormous potential energy of the strong field. Fission simply turns it into kinetic energy. There is also a small loss of actual mass in this process, about half a gram in the case of the first atomic bomb. Fusion between nuclear particles does the same, but releases even more of the enormous energy contained in the nuclear particles themselves. Thus there is a greater loss of mass in a fusion explosion, although this is a fairly simplified view. If we could control fusion in an atomic reactor power station, as we can with fission, we would have a vast new source of energy. The trouble is the vast amount of heat energy required to start the process. In a fusion or H (hydrogen) bomb a fission bomb is used to start the reaction, but that might be a bit drastic in a fusion power station. The energy released in the fission or fusion process is pure energy, primordial heat, which emerged in the Big Bang and subsequently created particles and forces.

A simple atom can be completely defined by a very complex mathematical equation, indicating that physics and mathematics are two sides of the same coin, and therefore also indicating that the universe could theoretically be defined mathematically. It's not difficult to see why this is so when you remember that everything in the universe is made of just 3 particles, electrons, protons and neutrons (and neutrinos, but we'll worry about them later).

A Bit of Chemistry

The study of the 92 natural elements that make up the universe and the natural world including all living things, in other words, that make up Everything, is part of physics as well as chemistry. They are listed in the Periodic Table of the Elements (Diagram no.7a).

Before we look at the Periodic Table let's look at fire again. According to the law of the Conservation of Energy, fire releases e-m light and heat energy without loss. It does the same with mass and there is a law of the Conservation of Mass (or atoms). In a fire the atoms (including electrons) in molecules move around violently forming other mostly gas molecules and in doing so release energy, but no mass or energy is lost, just passed on in a different state. The exceptions to this law are radioactivity, which does involve some very slight loss of mass, and fission and fusion in stars or atomic explosions. From this law it follows that the immense number of atoms in the hugely complex molecules of the organic matter of your body will redistribute themselves when you die, and carry on regardless. Similarly at some stage long before your time the more complex atoms in your body came from outer space, because all such atoms can only be produced by the tremendous heat and pressure in the core of stars. You are made of stardust. One could say "I am in the universe, and the universe is in me". In fact your body is made almost entirely of carbon, calcium, water and air, with about 1% of other elements. Although the Big Bang produced all known matter in the form of hydrogen (H), helium (He) and a little bit of lithium (Li), all the rest of the elements were made subsequently within stars which coalesced from the hydrogen, and were distributed around the universe by supernovae explosions.

In the Periodic Table the set numbers of electrons allowed in each energy level in an atom, starting with the lowest energy level nearest the nucleus, are 2, 8, 8, 18, 18, 32, 32 (going from top to bottom in Diagram no.7a). The last energy level of 32 is incomplete as we haven't discovered all possible elements as yet (all those after 92 are man-made). Each element is defined by the number of electrons or protons it has (the Atomic Number or Element Number). The e-m force involved in an atom's outer energy level or shell, if that shell is not complete, will always try to pull in other electrons from adjacent atoms in order to completely fill itself, so the e-m force is responsible for the science of Chemistry. Where electrons can be shared by atoms within an element, like germanium (Ge) or silicon (Si) for example, they form what is called a valency bond, and by sharing the electrons each atom completes its outer shell, and the whole forms a molecule. This is also how a crystalline (mineral) substance is formed. If the electrons are shared by atoms of dissimilar elements in order to fill the outer shell of the more complex element, then a compound molecule is formed, such as H_2O, two atoms of hydrogen combining with one atom of oxygen (O) in order to fill the outer shell of the oxygen atom. How curious that two gases can combine to form a liquid indispensible to life. Silicon and oxygen make glass, which has the remarkable quality of transparency. This type of chemical combining can also be destructive, as with rust or acids. Elements and compounds can have amazing and violently different properties. Everything in the universe, including organic entities, is made of the 92 elements, and compounds of them. Now let's look in more detail at the Periodic Table of the Elements (see Diagrams nos. 7a & 7b) from which you can check the H_2O combination.

The order of energy levels, also called periods, is from top to bottom. The number of electrons which is equal to the number of protons is the atomic number, which determines the chemical properties of the elements, and runs from left to right. Thus the elements in the right-hand column all have their energy levels or shells complete, and are called the noble, rare or inert gases. They do not form molecules. Carbon (C) has four electrons in the second energy level and six altogether. It is an element whose atoms can easily combine with its own atoms (as in a diamond, which is a crystal and hard, or graphite, which is soft) and also with those of other elements in attempting to fill its second (outer) shell. It is so versatile it has become the basis of organic matter, including you, because it can form long strings of molecules, such as are found in proteins.

The vertical columns or groups consist of elements that are related to each other. The group with elements 29, 47 and 79 in it is an example. These elements, copper, silver and gold, are fairly inert but mix well with each other.

The elements after 92 uranium (U), such as 94 plutonium (Pu), are all man-made and decay increasingly rapidly via radioactivity into elements of lower atomic number. Uranium itself may decay eventually to more or less inert lead but very, very slowly. Alchemists might have loved it, even though it isn't gold. The elements immediately preceding uranium in the table are also moderately radioactive.

Mercury (Hg) and bromine (Be) are liquids at room temperature and pressure. Hydrogen (H), nitrogen (N), oxygen (O), fluorine (F), chlorine (Cl) and the right-hand column are all gases at room temperature. All others are solids at room temperature and pressure. 104 – 112 are obscure radioactive man-made elements and are not shown. They are dangerous and last for a very short time before decaying into lower order elements. They are classed as Transitional Metals, and there are probably more to be created by adding successively more protons into the nucleus of uranium, thus turning one element into another, which is normally impossible. Temperature and pressure can be used to change elements between the three states of solid, liquid and gaseous, and also to help turn elements into compounds.

Isotopes are elements which have a lesser or more usually greater number of neutrons than they should. They may become unstable and decay back to the element. This decaying or radioactive process is the emission of neutrons or pairs of neutrons called alpha particles, and can be quite damaging to organisms. Helium nuclei can also be referred to as alpha particles. They have two neutrons and two protons. Beta particles are simply electrons. Gamma particles or rays are high frequency and thus high energy e-m radiation, deeply penetrating and also damaging. These last two can also be involved in radioactive decay. The rate of decay is measured by calculating the half-life of a radioactive element or isotope. Alpha, beta and gamma radiations also come from the fission and fusion processes, and thus also in the radiation from the sun, called the solar wind. Note that the atomic weight of an element, which is related to actual weight, is the number of its protons plus its normal number of neutrons, or plus the average number of neutrons in its most stable isotope if it is at all unstable.

Positive ions are atoms where an electron or electrons have been knocked off the outer shell so that the atoms are positively charged, as in the ionosphere, or ozone layer, the outer part of the earth's atmosphere, where it's due to the absorption of u-v radiation from the sun, and which is why in a complicated way (re-emission) the sky appears blue. If most of the u-v were not absorbed in this way, we'd all be fried. The ionosphere can reflect e-m radio waves in the HF region in order to achieve long distance communication, a form of communication that is now largely redundant, replaced mainly by satellite communication. There's also long distance communication by cable.

Negative ions are where an electron or electrons have been added to the outer shell. You too can create ions by rubbing a comb against your sleeve and using the resultant charge on the comb to attract a small piece of paper. This is an example of static electricity, or stored charge, where positive and negative charges, or ions, are held separately, as in a battery. However it can lead to a catastrophic discharge of electrons from one to the other, as in lightning. The Auroras Borealis and Australis are caused by the stream of mostly alpha particles from the sun's fusion process, the solar wind, interacting with the earth's magnetic field, without which we'd fry, yet again. This region is called the Van Allen belt. There is an 11 year sun cycle when such activity increases.

A mixture of elements or compounds which could combine into other compounds catastrophically easily, given the chance, together with a massive release of energy, is a possibly explosive mixture which sometimes just needs something to start it off. It's then like a fire, but all at once. A controlled reaction of this nature can fuel a rocket.

In a fire the atoms in the complex organic molecules of the substances involved, like wood, simply release e-m energy in the form of light and heat and also rearrange themselves, via the movement of electrons between atoms, and using the oxygen in air, becoming mostly molecules of gases with lower overall energy levels than in the original combustible materials, an example of entropy at work. But neither mass nor energy is lost, simply changed and, in the case of energy, radiated and conducted away, dispersed, but not lost. Oxygen is quite volatile as it seeks just two electrons to fill its second, outer energy shell. A fire is a complicated self-sustaining chemical reaction.

Your body is quite good at extracting energy from the food, drink and oxygen that we take in. It does this by very efficiently breaking down the molecular structure of what we take in, thus releasing that mysterious thing we call energy.

By analysing the intensity of the colours in the spectrum (Red, Orange, Yellow, Green, Blue, Indigo, Violet, the colours of the rainbow) which make up the white light from a star, it is possible to calculate the relative amounts of the elements in a star, since excited elements radiate different frequencies, and thus different colours. An element can also absorb the same colour. Of course the universe radiates across the whole spectrum of e-m radiation and studying such radiation (like x-rays) can tell us a lot about the universe.

The Electro-Magnetic Spectrum   Diagram no. 6

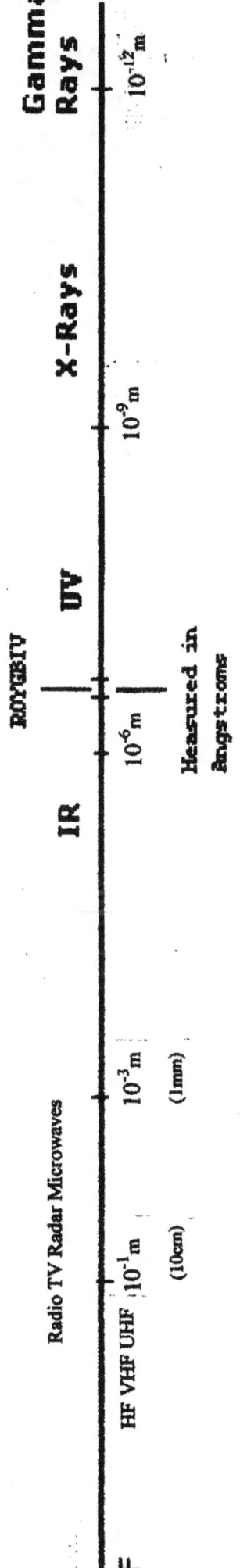

VLF   HF VHF UHF   Radio TV Radar Microwaves   IR   ROYGBIV   UV   X-Rays   Gamma Rays

10^{-1} m 10^{-3} m 10^{-6} m 10^{-9} m 10^{-12} m

(10cm)   (1mm)

Measured in Angstroms

Approx. Wavelength Decreasing ➔
(Frequency Increasing, Energy Increasing)

IR = Infra-red heat radiation
1 Angstrom = Diameter of an atom

ROYGBIV = Colours of the rainbow which
together make white light. I stands
for indigo. Red, green and blue are
the principal colours.

One Cycle of a Sine Wave λ = Wavelength

The Periodic Table of the Elements

Diagram no. 7a

KEY

Atomic number
Atomic Symbol
Atomic weight*

Each cell below is given as: Atomic number / Symbol / Atomic weight*

Period	Group I	Group II												Group III	Group IV	Group V	Group VI	Group VII	Group VIII
1	1 H 1.008																		2 He 4.003
2	3 Li 6.941	4 Be 9.012												5 B 10.81	6 C 12.01	7 N 14.01	8 O 16.00	9 F 19.00	10 Ne 20.18
3	11 Na 22.99	12 Mg 24.31												13 Al 26.98	14 Si 28.09	15 P 30.97	16 S 32.06	17 Cl 35.45	18 Ar 39.95
4	19 K 39.10	20 Ca 40.08	21 Sc 44.96	22 Ti 47.90	23 V 50.94	24 Cr 52.00	25 Mn 54.94	26 Fe 55.85	27 Co 58.93	28 Ni 58.71	29 Cu 63.55	30 Zn 65.38		31 Ga 69.72	32 Ge 72.59	33 As 74.92	34 Se 78.96	35 Br 79.90	36 Kr 83.80
5	37 Rb 85.47	38 Sr 87.62	39 Y 88.91	40 Zr 91.22	41 Nb 91.22	42 Mo 95.94	43 Tc (99)	44 Ru 101.1	45 Rh 102.9	46 Pd 106.4	47 Ag 107.9	48 Cd 112.4		49 In 114.8	50 Sn 118.7	51 Sb 121.8	52 Te 127.6	53 I 126.9	54 Xe 131.3
6	55 Cs 132.9	56 Ba 137.3	57 La 138.9	72 Hf 178.5	73 Ta 180.9	74 W 183.9	75 Re 186.2	76 Os 190.2	77 Ir 192.2	78 Pt 195.1	79 Au 197.0	80 Hg 200.6		81 Tl 204.4	82 Pb 207.2	83 Bi 209.0	84 Po (210)	85 At (210)	86 Rn (222)
7	87 Fr (223)	88 Ra 226.0	89 Ac (227)																

Lanthanide series

58 Ce 140.1	59 Pr 140.9	60 Nd 144.2	61 Pm (145)	62 Sm 150.4	63 Eu 152.0	64 Gd 157.3	65 Tb 158.9	66 Dy 162.5	67 Ho 164.9	68 Er 167.3	69 Tm 168.9	70 Yb 173.0	71 Lu 175.0

Actinide series

90 Th 232.0	91 Pa 231.0	92 U 238.0	93 Np 237.0	94 Pu (244)	95 Am (243)	96 Cm (247)	97 Bk (247)	98 Cf (251)	99 Es (254)	100 Fm (253)	101 Md (256)	102 No (254)	103 Lw (257)

Group I — Alkaline Metals

Group VIII — Noble or Rare Gases

Group VII — Halogens (Non-Metals)

Semi-Metals and Non-Metals

Transitional Metals

*The number in ( ) = Mass Number of the most stable isotope.

Metals are ductile and good conductors of heat and electricity

Names of the Principal Elements

Atomic No	Symbol	Name	Atomic No	Symbol	Name
1	H	Hydrogen	43	Tc	Technerium
2	He	Helium	44	Ru	Ruthenium
3	Li	Lithium	45	Rh	Rhodium
4	Be	Beryllium	46	Pd	Palladium
5	B	Boron	47	Ag	Silver
6	C	Carbon	48	Cd	Cadmium
7	N	Nitrogen	49	In	Indium
8	O	Oxygen	50	Sn	Tin
9	F	Fluoride	51	Sb	Antimony
10	Ne	Neon	52	Te	Tellurium
11	Na	Sodium	53	I	Iodine
12	Mg	Magnesium	54	Xe	Xenon
13	Al	Aluminium	55	Cs	Cesium
14	Si	Silicon	56	Ba	Barium
15	P	Phosphorus			
16	S	Sulphur	72	Hf	Hafnium
17	Cl	Chlorine	73	Ta	Tantalum
18	Ar	Argon	74	W	Tungsten
19	K	Potassium	75	Re	Rhenium
20	Ca	Calcium	76	Os	Osmium
21	Sc	Scandium	77	Ir	Iridium
22	Ti	Titanium	78	Pt	Platinum
23	V	Vanadium	79	Au	Gold
24	Cr	Chromium	80	Hg	Mercury
25	Mn	Manganese	81	Ti	Thalium
26	Fe	Iron	82	Pb	Lead
27	Co	Cobalt	83	Bi	Bismuth
28	Ni	Nickel	84	Po	Polonium
29	Cu	Copper	85	At	Astatine
30	Zn	Zinc	86	Rn	Radon
31	Ga	Gallium	87	Fr	Francium
32	Ge	Germanium	88	Ra	Radium
33	As	Arsenic	89	Ac	Actinium
34	Se	Selenium	90	Th	Thorium
35	Br	Bromine	91	Pa	Protactinium
36	Kr	Krypton	92	U	Uranium
37	Rb	Rubidium			
38	Sr	Strontium			
39	Y	Ytrium			
40	Zr	Zirconium			
41	Nb	Niobium			
42	Mo	Molybedenum			

Particles

I make no apology for delving into particles in some detail. It is only by a proper understanding of the basic structure of matter that the true nature of Reality will eventually be approached. You could skip this chapter and still capture the sense of this book, but I hope you don't. Although tricky and full of Lateral Thinking, it does give further insight into Reality. To start with, in conjunction with the table in Diagram no.8, I will simply list statements describing the micro-universe:-

The table shows all the standard particles and forces or fields. Dots denote a continuation of a similar pattern. 1/137 is to do with the charge on a proton. 2 up quarks and 1 down quark make a proton (charge 1/137). 2 down quarks and 1 up quark make a neutron (charge 0).

The names of the quarks are called flavours. Quarks also come in colours to do with how quarks are affected by gluons. Each quark comes in all 3 colours, red, green and blue.

Isospin or simply spin is the name for a characteristic of particles, which is best thought of as a sort of spin. Spin can be classified as ½, 1, 1½, or 2, mainly ½. SU (3) means Special Unitary Group (3 dimensions). Spin has only 2 dimensions…very odd. Both spin and colour are a form of charge, or coupled charge as the table calls it.

The figures give the approximate relative strengths of the appropriate forces on each particle. Note that G couples to or affects all energy inside the atom, but very lightly. Note the extremely weak effect it has on matter in the atom (LH column) compared to the other 3 forces.

Only the first and lightest of the 3 families (no.1) is common in ordinary matter. Families 2 and 3 decay rapidly into 1, and are highly energetic. They seem to be attempts to create matter which were non-viable, but are still with us.

There are 6 leptons (electrons, neutrinos etc.), 6 quarks and 6 bosons i.e. those shown as force particles plus the Higgs (see below). (666, does that remind you of anything?). The matter particles (leptons and quarks) are together known as fermions. Protons and neutrons are called hadrons or baryons.

Neutrinos come mostly from the sun, exist everywhere and are highly penetrating but not damaging. Untold millions are continually passing through you. They are unimaginably small, even compared to an atom. They have very, very little mass. They have spin but no electrical charge and can easily pass through an atom without hitting or disturbing anything. Neutrinos are very odd, and they don't seem to have a purpose. Their very rare collisions inside atoms are detected and studied in places like deep mine shafts, since only neutrinos can penetrate the earth that far down.

For all particles there exist anti-particles i.e. with the same mass but opposite charges, which would catastrophically annihilate each other by turning all their mass into energy if brought together, similar to the theoretical virtual particles that possibly create Dark Energy and the expansion of space. Examples of anti-particles are the positron (an anti-electron, i.e. with positive charge), the anti-proton (negative charge), squarks (opposite to quarks), photrinons (opposite to photons), and so on. Together with the other anti-particles, including some more obscure ones, such as those with opposite spins to the standard model, they form the super-symmetric table. Why exactly our universe uses our particles and not anti–particles is not known. It may be that there are universes where the anti-particles hold sway, or indeed totally different particles hold sway, and it is random as to which set a universe goes for. There are at least 150 particles. Some may just have been dead ends. Most particles are highly energetic and unstable and appear for very, very short times. They decay into one or more stable particles.

Particles can combine to form other particles, such as an electron plus a proton making a neutron (the charges cancel). It's what happens in a neutron star. Similarly a neutron can split into its two constituent particles.

Particles carrying the forces (bosons) are called virtual if the force or field is static i.e. mainly inside the atom, because such particles are effectively held within the associated matter particles, along with all sorts of other more fundamental particles. For instance the strong force bosons are effectively held within the quarks, and photons within electrons and quarks (see Diagram). G is static (normally) and its effect is mainly outside the atom. E-m can also be static outside the atom. All these 4 static fields carry potential (negative) energy, and their virtual bosons are presumably mass-less.

Virtual particles inside the matter particles are released, becoming non-virtual radiation outside the atom, if the matter particle is disturbed enough. It's another way of looking mainly at e-m radiation, since the other 3 radiations can be neglected. Thus a photon moving outside the atom i.e. released from within an electron, is not virtual, has kinetic energy but no mass, and can be regarded as a wave or as a particle.

Higgs bosons are virtual particles having mass, and they exist inside any matter particle thereby contributing mass to the particle. It's a bit of an oddity, for instance it is chargeless, including 0 spin. The Higgs boson was discovered in the LHC in the summer of 2012. Because it's a boson it implies the existence of a mass field, which sounds rather like G. Think of it like this. A particle is simply solidified pure energy, and pure energy of itself does not have mass. The Higgs boson supplies this aspect of reality, just as the other bosons supply theirs, enabling matter to exist in its finely balanced way. It seems to be a fundamental aspect of nature that at enormously high temperatures, as in the Big Bang, the energy in heat will convert into matter in this complicated way, and the temperature will drop. To detect the Higgs it was necessary to recreate the temperature conditions shortly after the Big Bang by colliding protons at enormous speed and therefore energy. Some of the energy making up a particle such as a proton, when released via such a collision, will form one or more Higgs emitted particles and any other appropriate particles or bosons.

From $E=mC^2$ it is clear that mass implies high energy and therefore a Higgs boson is also a high energy virtual particle. High energy emitted particles are very, very short-lived and the Higgs was difficult to detect because of this and also because of the enormously high temperature required. The Higgs mass field mentioned above must be added to the list of the other 4 atomic forces or fields. Shouldn't there be an anti-Higgs?

A fifth column on the Diagram representing the Higgs would be headed "Mass", and underneath "Higgs boson", underneath that "confined" (probably), then a blank (the Higgs is chargeless) and then figures and dots representing the degrees of interaction with the matter particles.

An odd thing is that if a photon had any hint of mass at all then, since it travels at C, its mass would theoretically become infinite – an impossibility, but of course it does have energy. Better to define a photon as pure energy, in that no Higgs bosons are present to give it mass.

Energy and mass get a bit mixed up in the micro-universe, as you may have gathered. Mass is measured in electron-volts in the micro-universe, which is actually a measure of energy. It's more convenient. Massive in the micro-universe means highly energetic, not necessarily big, so we must be careful how we use the word massive. A particle that is small can have a high mass and therefore high energy via the Higgs virtual particle or particles it contains.

Just as the G field provides weight to mass outside the atom, so it seems does the Higgs field provide mass to particles inside the atom.

Whilst not wanting to complicate things too much, we must not confuse mass outside the atom with weight. Weight is associated with G. G exists inside an atom, but it is so weak there as to be negligible. Consider the re-arrangement $m = E/C^2$. This implies that some forms of energy outside the atom can be equivalent to a very, very small mass (and could be converted into it were it not for the Law of Entropy). But mass has weight in a G field, so some forms of energy outside the atom have weight too, mainly heat energy but also motion. Thus a hot cup of tea weighs infinitesimally more than a cold one, and a moving object weighs slightly more than a stationary one, becoming infinite at C. You could say that the addition of weight due to heat is to do with the expansion of hot objects. From $E=mC^2$ it is clear that as the energy of either heat or motion increases, so does mass and therefore weight, C being constant. In the macro-world mass in general is a result of density – how tightly atoms can pack themselves together.

However in the warming up process of a liquid or a gas, the expansion of the warmer part means it weighs less and will therefore rise, and the colder part will sink, a process called convection, as in the world's oceans or atmosphere.

CERN in Switzerland operates the LHC, which everyone hopes will cast further light on reality. It is the largest machine in the world, measuring 27km in circumference, 8km in diameter and is a circular proton to proton collider buried deep underground. It is a type of cyclotron, a particle accelerator. At full speed a proton will circle the tube more than ten thousand times in a single second before colliding with one travelling in the opposite direction. This process creates a collision speed of nearly twice C, with the hope that the resulting massive high temperature fission/fusion will reveal more fundamental massive radiated short-lived particles from inside protons, including a radiated Higgs. Or in fact just the tracks or other evidence of these short-lived particles in the cloud chamber, which is simply a smallish chamber with a vapour cloud inside it. In order to "see" these particles it is necessary to create the massive heat energy of the collision in order to recreate a time close to the Big Bang when particles were still being formed.

Fermilab near Chicago houses another large cyclotron, called a Tevatron, but it is only one third as powerful as CERN.

Superstrings

The theory of Superstrings, which leads to Branes and M-theory (see below), is entirely theoretical and mathematical. The experimental energy required is far too big. It's a very powerful tool in resolving the difference between Quantum Mechanics (see below) and G, the main component of relativity, or you could say between the micro and the macro aspects of the universe. Such a solution is known as Quantum Gravity. The discovery of the mathematical nature of time may also help in solving Quantum Gravity. The Grand Unified Theory (GUT) would be the mathematical tool that would link together all the known forces, at a stretch including G, which together with Quantum Gravity should then give us the so-called Theory of Everything (TOE), if there can be such a thing. G seems to be a pivotal concept. What with G, M-theory, Dark Energy, Quantum Mechanics and everything, I suspect that the Theory of Everything may turn out to be a Theory of Some Things. We will just uncover deeper mysteries. But perhaps the LHC will take us deeper, at least.

Superstrings is a candidate for Quantum Gravity, but there are several versions, leading via Branes to M-theory (see next section) which is, just possibly, a candidate for the TOE, and predicts the existence of other universes. Superstring theory predicts the super-symmetry of particles mentioned earlier, and also black holes, but it also predicts 11 dimensions, consisting of our 3, plus time as a dimension, plus 7 others, curled up and existing at the superstring level of micro-matter. Our 3 became uncurled and expanded with the Big Bang to form our universe, whilst remaining curled up at the edges of the universe, as it were.

Superstrings themselves seem to be vibrating loops of energy which exist in the 3 plus 7 dimensions. Only complete or standing wave vibrations are allowed, meaning that the distance around each loop is an exact number of a particular wavelength. Each particular wavelength vibrating loop or string appears to correspond to a particular particle. In effect we are "seeing" the part of the loop in our 3 dimensions as a particle. There could also be straight strings, vibrating in the same standing wave way, whose ends we are "seeing" as particles, if that gives a better picture. Anything would help, so imagine a hosepipe, which represents a curled over dimension, with one loop of string wrapped around it. Seen from far enough away the string appears to be a point, a particle. That at least accounts for one extra dimension. Don't ask about the other 6!

Just for a bit of fun here is another Lateral Thought. Imagine a 3D person passing through a 2D world (flatland). A 2D being existing on that flatland world would see successive slices of the 3D person as she or he passes through the 2D plane, rather like what you see in one of those brain scan machines. Gruesome, I know, but it's only Lateral Thinking. Now imagine if you can a 4D person or object from a 4D universe, that is, one that has one further space dimension uncurled. Imagine this person or object passing through a 3D world. Alright, you can't. But if the object was a 4D sphere, wouldn't you first see a dot, then a very small 3D sphere, gradually getting bigger, and so on, on until the biggest sphere appeared, and thereafter a gradually diminishing 3D sphere, finally back to a dot, and then vanishing? Now how about if this other universe was more than 4D? Let's not go there.

A simple view of a 4D cube is found by first considering a square in 2D, which has 4 corners, then projecting it to a cube in 3D, which has 8 corners. Thence by simple logic we can deduce that a cube projected into 4D…a tesseract, must have 16 corners, although you can't imagine it. However it exists mathematically, and since all physics is mathematically based, it must be possible for 4D to exist in reality, and if 4D can exist, so can 5D and so on.

Let's speculate for a moment, using some lateral thinking. In our universe with our physics and dimensions and atomic structure and G, there is a natural limit to the small and large size of things, both inorganic and organic. In another universe with different parameters, and even possibly no atomic structure or G as we know it, then size could be a very different thing. Imagine, for instance, a vast living homogenous mass of some sort. Only energy would be the same. It's the realm of science fiction.

An interesting and increasingly accepted idea is that the reason the "shape" and "extent" of the universe appear so strange to us is because the universe actually exists in 4 space dimensions, but inside it you are only aware of 3. There are good reasons for this to be true, and it would explain much that seems weird to us about our universe. So when a 3D universe is mentioned in the text, remember, it could be a 4D one. It's a concept similar to a hologram.

Complex or imaginary number maths is a very simple tool employed in the sort of maths associated with superstrings. It's a maths of dimensions.

Branes and M-Theory

You must be getting fed up with all this Lateral Thinking. But inevitably as we near the end of our description of Reality it's going to increase, I'm afraid. Branes, also called Brane-worlds or M-branes (membranes), arise out of Superstring theory. Branes are postulated to represent other universes that exist in other dimensions "alongside" ours, yet undetectable by us. There are several versions, for instance P-branes (physicists humour again), leading on to what is known as M-theory. In essence this postulates the existence of a vast number of other universes which came into being at the same time as ours in the fraction of time after the Big Bang, and due to minute random quantum fluctuations in the dense energy ball emergent at that time. All these burgeoning universes underwent a catastrophic inflation just like ours, and some may be very like ours, some violently different from ours, with probably entirely different physics and dimensions if any, and many versions in between, most of which would not be viable and would not survive. They would have their own space-times and would be undetectable by us. The most probable ones would not have developed any matter or agglomerations of matter at all after inflation, but would have remained just a smoothly expanding ball of homogenous energy in space-time. They would not have developed the right physical constants necessary for matter to form. The least probable might have developed violent perturbations and would probably be non-viable. Ours would have a probability less than the most probable, but that's not to say that there might not be other universes similar to ours. This scenario implies a truly vast explosion of energy in the Big Bang, but it says nothing about black holes, or about the Big Bang itself or why the Big Bang occurred.

So whether any particular universe actually exists seems to depend on a Bell curve probability function (see Diagram no.10) which is, it seems, in accordance with the laws of Quantum Mechanics (see next section). Universes similar to ours developed only slight perturbations in the pre-inflation ball of energy, just like ours, probably with three dimensions and where the constants of their physics self-adjusted to a viable system. Very small changes in the constants of our physics, such as the coupled charges, would render our universe non-viable, but that is not to say that other viable systems could not exist. However we haven't observed them, so they exist only as a fuzzy mathematical probability as far as we're concerned (see The Role of the Observer later in this chapter). If indeed this scenario does turn out to be the Theory of Everything, how could it ever be proved? It would take a much bigger cyclotron than the LHC to get close to the vast energies and temperatures where superstrings might be revealed, which presumably were born when the early Big Bang was at those energies and temperatures. It could thence in some way confirm Branes and M-theory.

Neutrinos emanating from the LHC have been detected apparently travelling faster than C. If true, it throws the whole of our physics out of the window. However a possible explanation is that they momentarily pass into another "place", like another dimension, or another M-brane, or another universe, and then pop back into our universe, having apparently travelled faster than C. If so, this could perhaps confirm the existence of other "places", and possibly lead to the Theory of Everything, but there's a long way to go yet. The neutrino and neutralino (see Chapter 1 on Dark Matter) are super-symmetric and both are mysterious – could they be our link to other "places" or "elsewheres?",…other forms of Reality..?

Another explanation might be this. From $E=mC^2$ we can see that if, for the neutrino and neutralino, the relationship between energy and matter has a different value in some way from the rest of creation, then from the formula it follows that C will have a different value. So the neutrino and neutralino might be able to travel at a faster C than the rest of creation without violating basic physics. Implausible perhaps, but possible…?

Quantum Mechanics

It is possible Superstrings and Branes should come in this section, but it seems to make better sense to put them after Particles. You may think quantum mechanics or, if you like, quantum physics is going to be as obscure as superstrings and branes, yet electronics relies on quantum mechanics, it's why a transistor works. We wouldn't have the modern electronics industry without quantum mechanics, or rather without the phenomenon that quantum mechanics describes. It concerns the wave-particle duality that we mentioned earlier. When it was first mooted, physicists couldn't believe that anything so outrageously weird could lie at the very heart of physics. The experiment that confirms this bizarre behaviour is as follows. If particles, usually electrons (or photons) are directed at two parallel slits in a screen, the result on another screen on the other side of the first is an interference pattern, such as is normally formed by waves like light waves. But another interesting thing is that if you try to determine which slit particular electrons pass through, you can't without the pattern vanishing. It relates to the idea we've called the Uncertainty Principle, that is, if you know the position of an electron at a given instant, you can't know its path. If you set up two detectors to measure both path and position at any instant, one of them will not register anything. You could call the wave interference pattern on the screen a probability pattern, or which slit an electron passes through a probability function, and that's what quantum mechanics comes down to, probabilities. After all, particles are made of energy, and nobody knows exactly what energy is, so it's not surprising that it can do strange things.

The "Sum over Histories" theory says the same sort of thing in a different way. A particle travelling from one place to another has probabilities of taking particular paths, some highly probable, the most highly probable ones being more or less straight lines, and some highly improbable, and all paths in between, in other words virtually all possible paths. You can therefore create a probability picture for the path of the particle which is similar to the smoke ring idea we had for electron paths around a nucleus, except its linear rather than elliptical. However if you isolate the particle, then the picture collapses, called the collapse of the wave function. Until you isolate it, it's a wave and the best you can say about its path is the probability picture (see Diagrams nos.9 & 10). The particle then is acting purely as the energy which we know it's made of.

Another view of this phenomenon, and one that allows the transistor to work, is that an electron can appear on the other side of an electrical potential (or voltage) barrier, when it shouldn't be able to cross that barrier. It must have taken some very peculiar or highly improbable path to get there. Nonetheless a few electrons out of millions have made it, and that is enough for a transistor amplifier to work.

Another example of quantum mechanics or the Uncertainty Principle is the half-life of radioactivity. You can't possibly tell when a particular particle will be emitted - all you can say is that the highly probable average number of emissions in a given time will be such and such. This is how you calculate the half-life of an element. It is a highly probable statistical calculation.

Going back to the transistor principle, another view would be to say that you are seeing an effect without an apparent cause - although it has a very small probability….in other words a possible random event. Now the Big Bang was apparently a random event but perhaps with a certain probability of happening, and because it came from a micro-source, a singularity, that is why I called it a random quantum event. If it is possible for a simple electron to appear, apparently out of nowhere in the micro-universe inside a transistor, why not the Big Bang in the macro-universe? We cannot know more, although we can speculate.

More esoterically, one can use the transistor principle as an even closer analogy for what possibly happens in a singularity, as in a black hole or even a Big Bang. In a transistor a few electrons (i.e. concentrated kinetic energy) randomly appear on the other side of a potential energy barrier, and thence draw kinetic energy from the potential energy, enabling a transistor to amplify, similar to the outgoing kinetic energy from the singularity in a Big Bang, or a Big Bang on the other side of a back hole.

In the macro-world we are well used to apparently random events…perhaps an earthquake or volcanic eruption or meteor strike. Even in the organic world we have, for instance, sudden unpredictable mutations in DNA, indeed accidents of all kinds, and this will always be the case. They do have causes, but it might be very difficult to isolate some particular trigger, so much so as to be called random, although again with a certain probability of happening within say a certain time frame. This is as true of human activities as it is of Nature's. Perhaps this is the reflection in the macro-world of what is happening in the micro-world. There are theories that reflect this. Catastrophe Theory is about sudden apparently random changes or events, sometimes called singularities, like the moment when a flower suddenly opens. The Butterfly Effect Theory (or Chaos Theory) is about how one beat of a butterfly's wings can, through effect after effect building up, lead to a particular pattern of air currents, whereas if the beat hadn't occurred the eventual air current pattern would have been very different. It's a bit of an exaggeration, but you get the idea. It means that for instance the weather can only be predicted for a few days ahead. After that the pattern becomes chaotic. The simple self-replicating feedback maths of this theory, by the way, leads to fractal theory, if you happen to know about that. It underlies the way all natural complex systems form themselves, although specific patterns are unpredictable. Examples are the way branches on trees form themselves, and even how coastlines appear. The pattern of stars and even galaxies that we see also conforms to this theory.

As an aside it's worth noting here that, in certain cases and in a sense opposite to the above, a mathematical theory of large numbers can be used to predict future trends or events in large number groups, like population shifts, given certain starting conditions and parameters.

To sum up, we have looked at two aspects of the quantum mechanical Uncertainty Principle that we can observe – firstly that energy radiation can appear as a wave or as a string of particles, which is linked to the fact that you cannot know both the path and the position of a particle, and secondly that you cannot predict when or whether a particular event in the micro-universe will occur, only probability functions apply. This last aspect may also apply to apparently random events in the cosmological macro-universe, in particular that Bell probability curves apply, including for the Big Bang itself. Let's therefore take events in order:-

1. The Big Bang itself was an apparently random quantum event and thus the probability of the existence of other Big Bangs, many being non-viable, possibly spread over time (which we can only imagine as being in our sense of the word time) must be considered as a Bell curve.

2. In the case of there being only one Big Bang, under the heading Branes and M-theory we stated that the existence of a vast number of M-theory universes may well follow a Bell probability curve.

3. In our universe, and possibly in others, in the super-symmetric model of particles many particles only exist for a minute fraction of time, others longer, and our No.1 family of quarks, electrons and neutrinos for an immensely long time. Thus the probable existence of particular particles at any given time should follow the Bell curve, our family perhaps being the most or nearly the most probable. Also perhaps the number of dimensions or expanded dimensions that came into being in the first place might also follow a Bell curve.

4. The formation of galactic black holes, formed by quantum rents in the expanding fabric of space-time after inflation, and thence the formation of galaxies and galaxy clusters, may also follow a Bell probability curve, both in the viability of emerging black holes and in time. Thus there may have been a brief era when black holes were more likely to form.

5. In the final chapter we consider the possibility among others of a multi-universe where individual universes are formed on the "other side" of galactic black holes. If the formation of the black holes themselves follows a Bell probability curve, so must the existence of universes spawned or connected by black holes.

The fuzziness associated with the non-observing of quantum events (see The Role of the Observer below) would also have a role in all these events. Overall we can conclude that Reality is governed by quantum theory, although in our universe we can only directly perceive its effects, and then only darkly, in our micro-universe quantum theory, and in our macro-universe probability theory. What ancient philosopher could ever have imagined the true uncertain nature of Reality, and thence the probable existence of other universes?

The Role of the Observer

This is an aspect of quantum mechanics following on from the Uncertainty Principle. If you attempt to measure something in the micro-world you affect what you are measuring, or you can say the presence of an observer determines the outcome, as in the slit experiment. This leads us to a very famous thought experiment, Schrödinger's Cat (see Diagram no.11). The cat is placed inside a closed opaque box in which there is a flask of lethal nerve gas. A weakly radioactive material is arranged in such a way that if it emits a particle (the emission of only one particle is a random event), a hammer drops and smashes the flask, killing the cat. Until you open the box you cannot know the outcome. In micro-world terms, the curious thing is the cat is both alive and dead until you open the box. It's a very good thought experiment in that it sums things up quite nicely. It illustrates the three aspects of what we've been getting at. Firstly, dual quantum reality i.e. both states existing at the same time (like wave-particle duality). Secondly, whether a random event takes place (that is, probabilities) and thirdly, the role of the observer determining the outcome. We can sum all this up as the fuzziness of non-observed quantum reality.

Let's go back for a moment, much as I hate to, to the Many Worlds theory (see Chapter 4 on Time Travel). Because we don't know how events, and even micro-world events, will affect our future, then you could say the future is fuzzy, but fuzzy around the most probable outcomes. This is, I think, a somewhat better picture of the indeterminate future than the Many Worlds interpretation. Until the future is observed (as time passes) it's not fixed. But it follows (perhaps) that in a strange and fuzzy way the future is out there, in a strange dimension of time (you could call it imaginary time), a mass of probabilities and possible random events, waiting to be fixed by being observed. A strange concept, but who knows?

If you've not reached saturation point, now consider freewill. Is it really free will? Or is it just the most probable outcome of all the random quantum activity which must be happening at micro-level in the thought and memory process parts of the brain, when prompted by a human (or indeed animal) motive? Even so, it's a form of freewill. There's a book called the Dice Man, in which a man determines his actions by the throw of a dice…certainly no probabilities or apparent freewill there then.

What about dreams? Are they the result of quantum activity in the sleeping brain? These are just some of the peculiar ideas that come out of quantum theory. Here's another, and it's fundamental. Could you have a universe where the micro-world is such that there is no uncertainty, no freewill, every event determined in advance? It wouldn't be much fun. There would be no probabilities. Given enough information, the future would be entirely predictable. The favourite would always win.

Quantum Entanglement occurs when you separate a pair of linked particles. When the spin of one is altered, the spin of the other, although possibly miles away, alters at exactly the same time. No one knows how they can do this. It's an action that is faster than C, so it must be to do with a minute instantaneous warp in an unknown field (or possibly dimension) that joins them, as with G in the disappearing sun thought experiment.

Quantum Computing is computing using micro-level components. Would a vastly complex Artificial Intelligence quantum computer dream due to random excitations in its memory circuits?

The micro-world is very strange. But you can deduce from this chapter that it is fundamental to the nature of Reality. It has provided clues to the nature of the multi-universe. The more we can discover about it, perhaps from the LHC, the more we will know about "how" our Reality came into being, but perhaps not "why".

Forces of Nature, Types of Matter and their Couplings in the Standard Model.

Force:	Gravity	Electro-magnetism	Strong Force	Weak Force
Force Particles (bosons):	gravitons	photons	8 gluons	$W^{\pm}$ and Z^0 bosons
Radiation:	gravitational waves	light etc.	confined	short range
Coupled charge:	all energy	electrical charge	SU(3) color (R,G,B)	SU(2) weak isospin
Matter Particles:				
up quark	$10^{-38} \times \left(\dfrac{\text{energy}}{m_{\text{proton}} c^2} \right)^2$	$\dfrac{2}{3} \times \dfrac{1}{137}$	10^{-1}	10^{-2}
down quark	•	$-\dfrac{1}{3} \times \dfrac{1}{137}$	•	•
electron	•	$-1 \times \dfrac{1}{137}$		•
electron neutrino	•			•
charmed quark	•	•	•	•
strange quark	•	•	•	•
muon	•	•		•
muon neutrino	•			•
top quark	•	•	•	•
bottom quark	•	•	•	•
tau lepton	•	•		•
tau neutrino	•			•
Phenomena:	stars, solar system, galaxies, universe	visible light, atoms, molecules, chemistry, biology	atomic nuclei, nuclear energy	radioactive decay, element formation

Illustrating Probabilities

An Amusement Arcade Machine

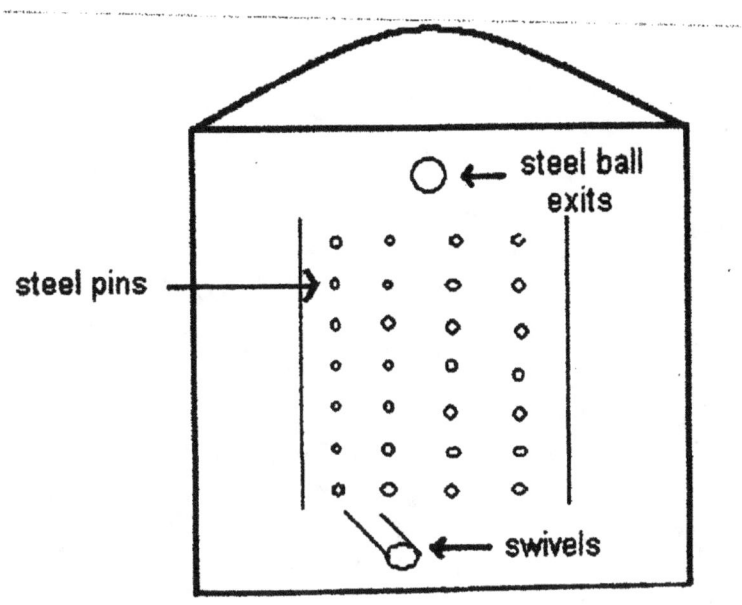

Ball caught returns penny

What is the probability of a second ball following precisely the same path as the first ball? It's a remote probability. Why? Because although they have the same starting point, or initial conditions, minute differences in physical conditions due to random effects mean the balls will almost certainly follow different paths. Now assume the machine is tilted backwards. The probability of the balls following the same path increases because the balls have less energy and bounce about less (also their paths are less chaotic and therefore exhibit lower entropy). This is an example of how quantum mechanics relates to the macro-world i.e. probabilities.

Probably the most probable thing happens, but there is an outside chance of an improbable thing happening.

More on Probabilities

Bell Curves

The Bell Curve is the standard probability curve. Dots show the number in each sample plotted against statistical probability.

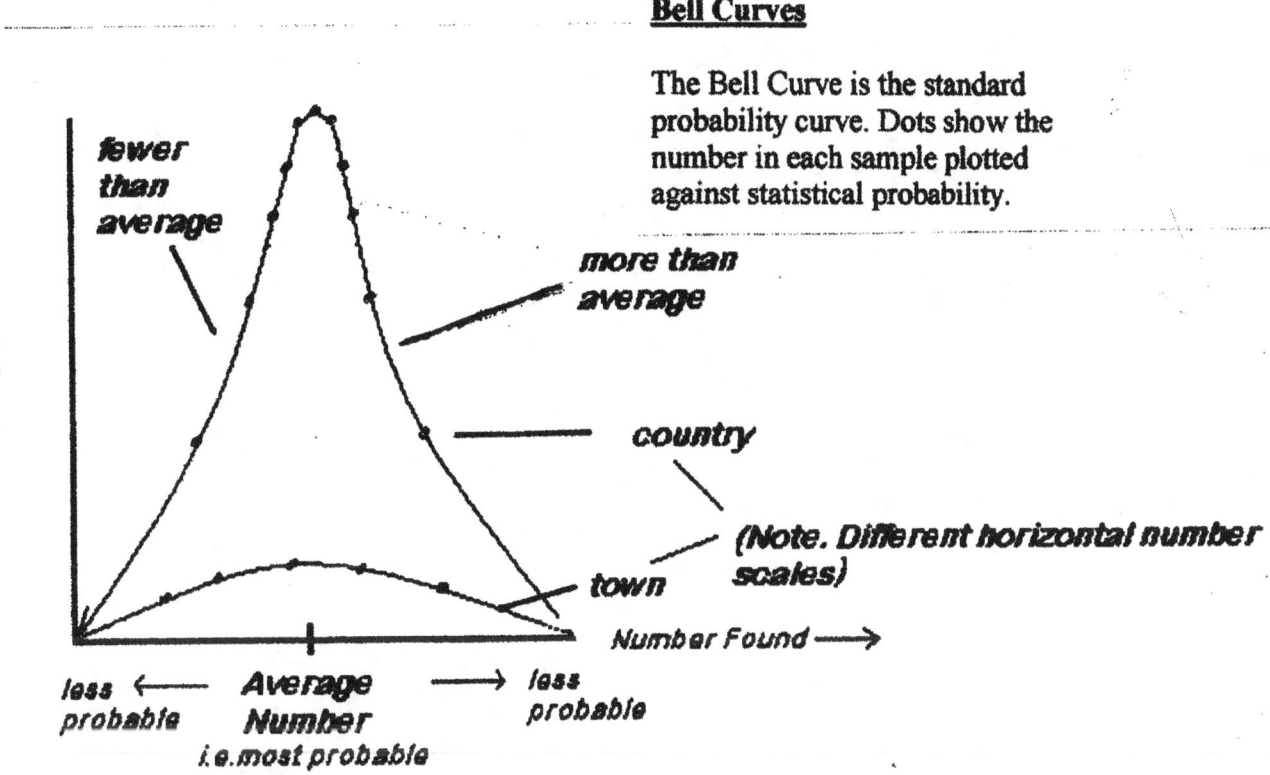

Graph of Number of Horse-riders per 100 of the Population

In the case of the steel ball falling through steel pins (Diagram No. 9) the horizontal scale above could represent, say, the number of balls following the same path per hundred runs, and the two graphs would be the results for the angle the board is positioned at, the upper graph being the result when the board is at a shallower angle i.e. the ball drops more slowly.

Schrödinger's Cat

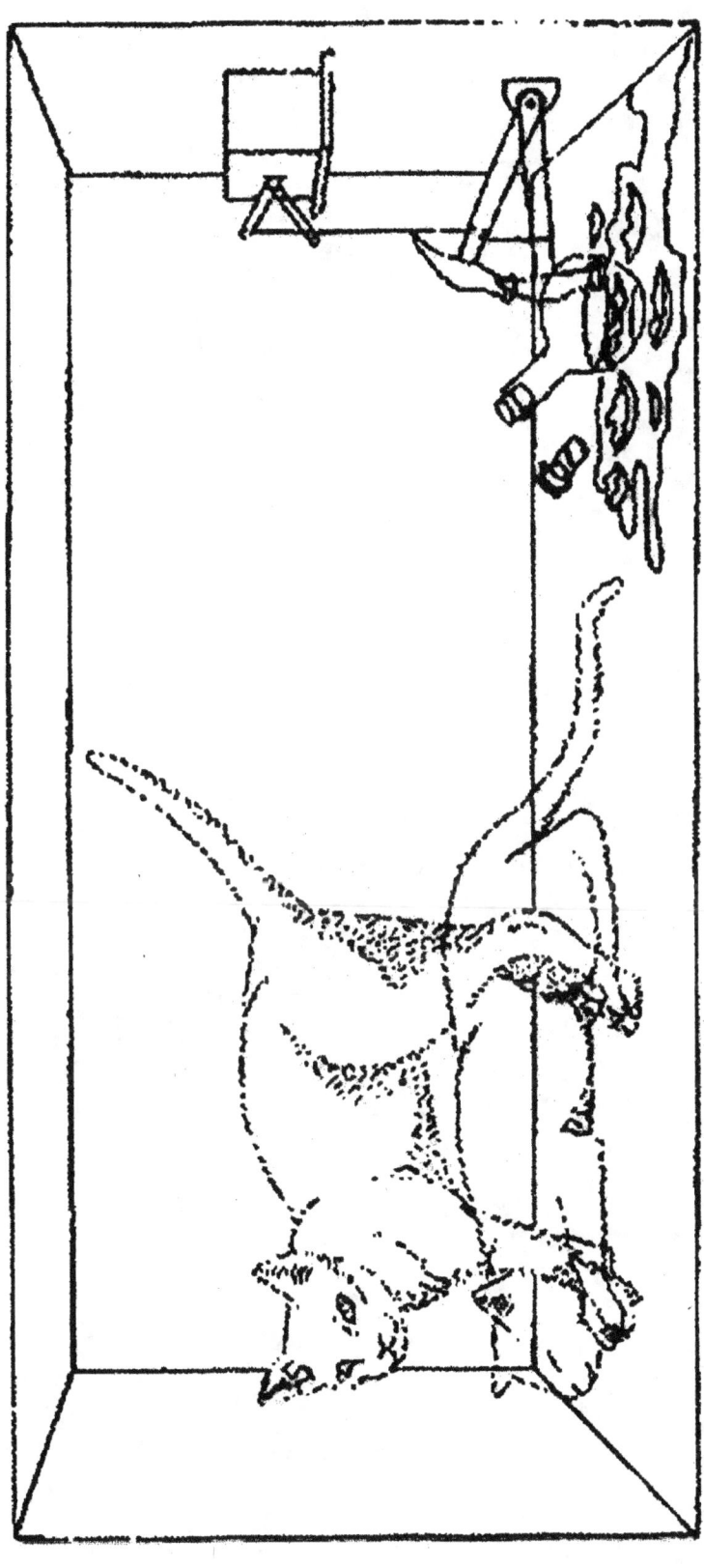

Schrödinger's Cat illustrates dual quantum reality, probabilities and the role of the observer. Ideally one could compute the probability of life or death (or at least the statistical probability) though that would not be of much use to the cat. We can sum this up as the fuzziness of non-observed quantum reality.

Chapter 7
The Multi-Universe
(or Multiverse)

We have mentioned the Anthropic Principle and the Goldilocks Effect. There's also a Weak Anthropic Principle (WAP) and a Strong Anthropic Principle (SAP), both of which we needn't worry about in any detail, and of course someone has dreamt up a Completely Ridiculous Anthropic Principle (....). More seriously, the Goldilocks Effect (not the Goldilocks Zone) is one way of explaining why things are just right for our universe to exist, in other words that we are one of many attempts, a lot of which fail. So the idea is that only in a few universes does the micro-world naturally sort itself out as it comes into being. Any answer, if there is one, may be indicated by the Theory of Everything, if we finally solve it. As just one example of the fine tuning of our universe, if G were any weaker we would already have recessed into total entropy, or if it were any stronger, collapsed into a Big Crunch.

There is the universal principle (Chaos Theory) that states that highly complex systems can become self-regulating, via the mechanism of feedback (reinforcement of what works). Examples include Evolution, the Gaia hypothesis of the globally interdependent ecology of the earth, and possibly the formation of the super-symmetry of inorganic particles and their values. It's the survival of the fittest or "trial and error" principle. This may also apply to the life and death of emerging universes, in that only those developing viable dimensions and physics are able to survive, and thence perhaps give birth to other universes via black holes. I see no reason why there should not be a finite number of such universes, possibly being born and dying according to a law of the Conservation of Cosmic Energy. Universes which are considerably smaller than ours could be viable, although probably only the vast galactic black holes in our larger galaxies could produce sizeable viable universes on their other sides. Overall one can say that certain cosmological events can be looked at from the point of view of Bell probability curves, Goldilocks Effects or Chaos Theory.

Perhaps the simplest multiverse scenario is to regard our universe as just one of very many universes resulting from one Big Bang, and perhaps with our black holes feeding into them, and theirs feeding into us and each other, accounting for our Dark Energy. This (without the black hole feeds) is the M-theory scenario. (There could be more than one Big Bang, possibly resulting in similar multiverses).

This scenario may, incredibly, be testable. If there do exist a vast number of universes, they may occasionally collide. It may therefore be possible to look for traces of such a collision in the CMB of our universe. More realistically, further discoveries at the LHC as to whether the super-symmetric table of particles is incomplete in this universe will point to the existence of the multiverse, where the missing particles would exist.

None of the above may be true. Our Big Bang may have uniquely, and randomly, occurred out of Nothing, and black holes may lead into Nothing. M-theory universes may simply be mathematical niceties. Our universe may be all there is, but even so there is enough of it to keep us wondering about it, perhaps for another few thousand years if we are lucky.

But if not, then in this scenario there is no explanation of an original Big Bang (or indeed what was "before" it) except as a random event. Cyclic theories like the Big Bounce get us no further forward, and offer no multiverse scenarios. One way of looking at an original Big Bang and what was "before" is as follows. The universe is made of energy in various forms. Even space is a form of energy. But energy itself has no form. It can be compressed in on itself as in a black hole singularity and apparently disappear, with positive (kinetic) and negative (potential) versions seeming to cancel out, or better expressed as kinetic energy being absorbed by potential energy, but then perhaps suffer a metamorphosis and re-appear as a non-original Big Bang on the other side of the black hole, separating out into positive and negative versions again. Therefore perhaps it follows that what exists before an original Big Bang is simply an infinity of potential energy, but with a certain probability of a Big Bang happening. We are talking about a cosmic ubiquitous potential energy that can exist in various forms of Reality, if (apparently) randomly triggered to do so. It is the substance of the multiverse, the cosmos. Note that entropy is maximum at a black hole singularity, but also becomes maximum when a universe ends in universal photons. Perhaps therefore maximum entropy is effectively the same thing as potential energy.

Just as we have the law of the Conservation of Energy and the law of Entropy, so there must be, one would think, an unknown third law which states that over the short term in viable universes energy will transform itself into as complicated a state as possible. Thus after the Big Bang, in our universe we have the creation of the atom, then the fledgling universe leading to 92 elements and eventually organic matter. Then entropy takes over.

Perhaps it follows that there is also a law which implies that potential energy will always try to become kinetic energy (and thence vice versa, which would be the law of Entropy). We stated in Chapter 5 that potential energy is simply kinetic energy waiting to happen. It's there in the name really. Suppose then the pressure to change in the cosmic ubiquitous potential energy builds up until it becomes irresistible. So a Big Bang would be a randomly triggered catastrophic way for the change to happen, perhaps subject to a Bell probability curve, something like a volcano. Then our unknown third law would apply, leading perhaps to a multiverse, then entropy in the individual universes, leading ultimately to only photons existing (experiencing no time) in any one universe, which would eventually over aeons simply degrade back to a pure or potential energy "foam". This is the M-theory scenario together with an explanation of the Big Bang.

But, if not, then still we have the problem of the original Big Bang. Perhaps therefore we could picture a multiverse like this. A boundless mass of ubiquitous cosmic potential energy "foam", inside which universe bubbles of different sizes randomly (the Uncertainty Principle) grow (Big Bangs and inflation), or disappear quickly, possibly inter-link with themselves and with the foam (black holes), and are then eventually re-absorbed (maximum entropy), going on for ever (except that "for ever" has no meaning in the "foam". It is timeless. Nor does it exist in any sort of space as we know it). That's as good a picture as any, as an alternative to the Big Bang and M-theory scenario.

An intriguing Lateral Thought here is that the singularities in the black holes linking the universe bubbles in the potential energy "foam" actually draw in vast amounts of potential energy from the "foam" before emerging as Big Bangs on their other sides. This is not unlike what happens on a much tinier scale in a transistor (see Quantum Mechanics). It's the Uncertainty Principle at work, and could mean that the universe bubbles in the "foam" are not too dissimilar to each other in energy content and therefore size. Moreover the incoming energy (Dark Energy) into a universe bubble would replace the outgoing energy lost via its black holes.

We said at the beginning of this book that the big question is "why does the universe (or the multiverse, or perhaps what was before the multiverse) exist?" to which there would seem to be no answer. But perhaps there is one after all, based on the weird randomness of the micro-universe, which is fundamental to reality. And that answer is itself a question, the ultimate question, "Why Not?"

And that is the end of "Cosmology in a Nutshell". Although complicated, I hope this book has helped you gain an insight into the immense mystery of creation. More and more of this mystery will be revealed as time goes on. We will come to understand many of the questions explored in this book, theories will come together and simplify, which is a sign in itself of truth in science and cosmology, but whether we will get a more definitive answer other than "Why Not?" I beg leave to doubt.

www.ingramcontent.com/pod-product-compliance
Lightning Source LLC
Chambersburg PA
CBHW081056170526

45166CB00006B/2086